Sustainability and Perspectives of Edible Insect Rearing and Utilization of Their Products and Byproducts

Sustainability and Perspectives of Edible Insect Rearing and Utilization of Their Products and Byproducts

Guest Editor

Simona Errico

MDPI

Basel • Beijing • Wuhan • Barcelona • Belgrade • Novi Sad • Cluj • Manchester

Guest Editor
Simona Errico
Laboratory of Regenerative
Circular Bioeconomy
Department for Sustainability
ENEA—Trisaia Research
Centre
Rotondella
Italy

Editorial Office
MDPI AG
Grosspeteranlage 5
4052 Basel, Switzerland

This is a reprint of the Special Issue, published open access by the journal *Agriculture* (ISSN 2077-0472), freely accessible at: https://www.mdpi.com/journal/agriculture/special_issues/PID291THC3.

For citation purposes, cite each article independently as indicated on the article page online and as indicated below:

Lastname, A.A.; Lastname, B.B. Article Title. *Journal Name* **Year**, *Volume Number*, Page Range.

ISBN 978-3-7258-5647-3 (Hbk)
ISBN 978-3-7258-5648-0 (PDF)
https://doi.org/10.3390/books978-3-7258-5648-0

Contents

About the Editor

Simona Errico

Simona Errico works at the Regenerative Circular Bioeconomy Laboratory, Sustainable Agri-Food Systems Division, Department of Sustainability at ENEA, Trisaia Research Centre, Rotondella (MT), Italy. She is a biologist who graduated from the University of Naples Federico II. Since 2003, she has been a researcher at ENEA—National Agency for New Technologies, Energy and Sustainable Economic Development—where, over the years, she has worked on various topics, from molecular biology techniques for studying the expression and levels of fatty acid desaturase in olive trees to monitoring the risk assessment of GMOs in protected areas and agroecosystems. Since 2015, she has been involved in the rearing of edible insects and their use as alternative protein sources. She is a panel leader in sensory analysis studies of functional foods and agricultural products obtained using alternative ingredients or innovative production and preservation processes. Since 2024, she has been a senior researcher at ENEA, Trisaia. She has participated in thirty national and international projects, as well as numerous conferences, and she has authored more than twenty publications, in addition to teaching several professional courses. She has also been involved in activities with schools and in the drafting of safety documents.

Preface

In recent years, worldwide interest in protein sources which can be used for human nutrition has focused attention on edible insects. The rearing conditions, the utilizable byproducts and even the insects themselves that are considered edible are not the same the world over. With this Special Issue, we wanted to gather scientific articles of international standing that would give an overview as complete as possible of the situation in the world regarding this topic. The result was the publication of relevant manuscripts (articles and reviews) mainly on three edible insects: Silkworm (*Bombyx mori*), Black Soldier Fly (*Hermetia illucens*), and Mealworm (*Tenebrio molitor*).

Simona Errico
Guest Editor

agriculture

MDPI

Editorial

Sustainability and Perspectives of Edible Insect Rearing and Utilization of Their Products and Byproducts

Simona Errico

Laboratory "Circular Regenerative Bioeconomy", Division "Sustainable Agri-Food Systems", Department "Sustainability", ENEA—Trisaia Research Centre, SS Jonica 106, Km 419+500, I-75026 Rotondella, Matera, Italy; simona.errico@enea.it

This Special Issue, titled "Sustainability and Perspectives of Edible Insect Rearing and Utilization of Their Products and Byproducts", aimed to gather high-quality scientific contributions suggesting innovative solutions for rearing edible insects and new perspectives on the use of rearing products and by-products. The prediction that the world's population will increase to 9 billion by 2050, but probably even earlier, resulting in an ever-increasing demand for protein, makes it necessary to find alternative protein sources [1].

Edible insect rearing can be useful in this context. However, it must be sustainable to avoid further burdening the environment and climate. This can be achieved by valorising waste and by-products for insect feed. Europe has only recently started using insects as human food, though insects have been bred for food purposes in other countries for a long time [2].

This Special Issue comprises five scientific studies and three reviews.

The first of those published reviews evaluates the sustainability and economic opportunity represented in one European country, Romania, by the rearing of the silkworm (*Bombyx mori*), which has been proposed for human consumption but is not yet approved by the European Commission. The authors analyse possible improvements in rearing, the crucial importance of preserving the mulberry tree habitat and the need to also use the by-products of rearing to make it sustainable. However, they suggest that further studies are needed on the nutritional qualities and possible health effects of silkworm larvae with a view of their potential use for human nutrition [3].

The second review analyses *Hermetia illucens* rearing, mainly by assessing its demand regarding nutrition, potential and performance measures. This contribution emphasises the difficulty of translating gained knowledge to industrial applications mainly due to the considerable differences in by-products used and the biotic and abiotic variables considered by previous studies, but it also helps define the range of variables to be tested in the future [4].

The last of the three reviews describes the valorisation of spent coffee grounds to rear black soldier fly (*Hermetia illucens*) larvae and illustrates the results of a bibliometric study, i.e., the statistical analysis of books, articles or other publications. This contribution explains how an adequate diet composition is indispensable since coffee contains caffeine as an alkaloid that hampers the growth of certain plants and *Hermetia illucens* larvae themselves. Furthermore, it analyses the possibility and limitations of using *Hermetia illucens* larvae fed in this way in aquaculture and fisheries [5].

Among the studies, Suraporn et al. propose using waste from Thai silkworm rears with the perspectives of both sustainability and economic benefit. Specifically, the group isolated lactic acid bacteria from the faeces of a local strain of *Bombyx mori* and obtained five extracts, all of which showed bile salt hydrolase activity, the ability to adhere to epithelial cells and remarkable probiotic properties, making them potential candidates for use as probiotics in feed [6].

Of the other four studies, two focus on *Hermetia illucens* rearing and two on mealworm (*Tenebrio molitor*).

The two studies focusing on *Hermetia illucens* larvae, one from Canada and one from Italy, explore their uses in areas other than human food because Canada has yet to implement specific regulations on entomophagy and Europe has not yet approved *Hermetia illucens* larvae for human consumption.

Bejaei and Cheng studied the effects of including dried *Hermetia illucens* larvae in the diets of free-range laying hens on production efficiency, food safety, blood metabolites and health. In general, their use had no negative effects on hen health and welfare; however, the authors emphasised the need for further studies to determine more efficient proportions of dried *Hermetia illucens* larvae in the diets of laying hens [7].

De Santis et al., on the other hand, proposed a multidisciplinary approach for developing a supply chain in the conversion of agro-food waste biomass mediated by *Hermetia illucens* larvae. The approach covers the entire process, from rearing to the exploitation of by-products. In their study, the heterogeneity of fruit and vegetable waste from large-scale distribution did not appear to be a limiting factor in standardising production in an insect rear and in organising the bioconversion process on an industrial scale. Furthermore, the researchers noted the possibility of using compost obtained from *Hermetia illucens* rear residues as a soil conditioner and as a partial replacement for synthetic fertiliser [8].

The other two studies address issues related to rearing *Tenebrio molitor*, the only insect discussed in this Special Issue that is presently approved by the European Commission.

The study by Errico et al. shows the advantages of using prickly pear (*Opuntia ficus indica*) cladodes as a source of water and nutrients for the small- and medium-scale rearing of *Tenebrio molitor* larvae, especially in the Mediterranean area, where this cactus is widespread, but its cladodes are rarely consumed. *Opuntia ficus indica* maintains its overall qualities longer than apple and potato while remaining palatable to mealworms, which allows for less frequent feeding. The possibility to utilise residual cladodes from pruning and the fact that it does not need to be stored at low temperatures makes Opuntia *ficus indica* an advantageous alternative water source that could further reduce the ecological footprint of rearing *Tenebrio molitor* at the small-to-medium production level [9].

Vrontaki et al. investigated the effects of using agricultural by-products in the large-scale rearing of *Tenebrio molitor* larvae on larval growth and body composition. In particular, the authors emphasise the benefit of incorporating locally abundant oat by-products, spent brewer's grains and rice bran into *Tenebrio molitor* diets. The work shows, on the one hand, the great advantage of this large-scale utilisation of *Tenebrio molitor* rears, and on the other hand, how it is not possible to directly apply the protocol and the results obtained to any by-product due to the unique characteristics of each matrix and the possible need for transforming the by-product before it is included in the *Tenebrio molitor* diet [10].

It is clear from these scientific contributions that the issue of sustainability in protein production is very much on the minds of researchers globally. Although dealing with very different specific topics, all of the studies aim to achieve the valorisation of by-products and agrifood waste (oat by-products, spent brewer's grains, rice bran, and prickly pear) and the use of rearing by-products in different sectors (using isolates of lactic acid bacteria from faeces as probiotics in animal nutrition, frass as a soil conditioner for vegetable crops, etc.).

Overall, these contributions provide valuable insights for new research, demonstrating the current relevance of this topic and its significance to researchers. They invite further studies to provide a comprehensive answer to the questions posed by society, politics, and consumers.

Acknowledgments: I would like to thank the Agriculture Editorial Office for the opportunity to be Guest Editor of this Special Issue, all the authors who sent in their contributions and all the re-viewers who provided valuable advice to improve the contributions. Acknowledgements are also due to my colleagues Alessandra Verardi for producing the graphical abstract and Paola Sangiorgio for her precious help.

Conflicts of Interest: The author declares no conflict of interest.

References

1. United Nations Department of Economic and Social Affairs, Population Division (2024). World Population Prospects 2024: Summary of Results (UN DESA/POP/2024/TR/NO. 9). United Nations—Department of Economic and Social Affairs Population Division. World Population Prospects 2024. Available online: https://desapublications.un.org/publications/world-population-prospects-2024-summary-results (accessed on 30 October 2024).
2. Omuse, E.R.; Tonnang, H.E.Z.; Yusuf, A.A.; Machekano, H.; Egonyu, J.P.; Kimathi, E.; Mohamed, S.F.; Kassie, M.; Subramanian, S.; Onditi, J.; et al. The global atlas of edible insects: Analysis of diversity and commonality contributing to food systems and sustainability. *Sci. Rep.* **2024**, *14*, 5045. [CrossRef] [PubMed] [PubMed Central]
3. Hăbeanu, M.; Gheorghe, A.; Mihalcea, T. Silkworm *Bombyx mori*—Sustainability and Economic Opportunity, Particularly for Romania. *Agriculture* **2023**, *13*, 1209. [CrossRef]
4. Nayak, A.; Rühl, M.; Klüber, P. *Hermetia illucens* (Diptera: Stratiomyidae): Need, Potentiality, and Performance Measures. *Agriculture* **2024**, *14*, 8. [CrossRef]
5. Hutabarat, D.J.C.; Mangindaan, D. Cultivation of Black Soldier Fly (*Hermetia illucens*) Larvae for the Valorization of Spent Coffee Ground: A Systematic Review and Bibliometric Study. *Agriculture* **2024**, *14*, 205. [CrossRef]
6. Suraporn, S.; Cansee, S.; Hupfauf, S.; Klammsteiner, T. Lactic Acid Bacteria from *Bombyx mori* Frass: Probiotic Properties and Antagonistic Activities. *Agriculture* **2024**, *14*, 924. [CrossRef]
7. Bejaei, M.; Cheng, K.M. Inclusion of Dried Black Soldier Fly Larvae in Free-Range Laying Hen Diets: Effects on Production Efficiency, Feed Safety, Blood Metabolites, and Hen Health. *Agriculture* **2024**, *14*, 31. [CrossRef]
8. De Santis, E.; de Iudicibus, A.; Lecce, F.; De Mei, M.; Petrazzuolo, F.; Del Giudice, A.; Carnevale, M.; Gallucci, F.; Beni, C.; Assirelli, A.; et al. A Multidisciplinary Approach for the Development of a Supply Chain in Biomass Conversion of Agrifood Waste Mediated by Larvae of *Hermetia illucens* L.: From Rearing to By-Product Exploitation. *Agriculture* **2024**, *14*, 1010. [CrossRef]
9. Errico, S.; Sangiorgio, P.; Moliterni, S.; Verardi, A.; Spagnoletta, A.; Dimatteo, S.; Baldacchino, F. Prickly Pear Cladodes as an Alternative Source of Water in Small and Medium-Scale Yellow Mealworm Rearing. *Agriculture* **2023**, *13*, 1435. [CrossRef]
10. Vrontaki, M.; Adamaki-Sotiraki, C.; Rumbos, C.I.; Anastasiadis, A.; Athanassiou, C.G. Bridging the Gap: Scaling Up the Sustainable Production of the Yellow Mealworm with Agricultural By-Products—Insights into Larval Growth and Body Composition. *Agriculture* **2024**, *14*, 520. [CrossRef]

agriculture

MDPI

Review

Silkworm *Bombyx mori*—Sustainability and Economic Opportunity, Particularly for Romania

Mihaela Hăbeanu, Anca Gheorghe * and Teodor Mihalcea

Research Station for Sericulture Baneasa, 013685 Bucharest, Romania; mihaela.habeanu@scsbaneasa.ro (M.H.)
* Correspondence: anca.gheorghe@scsbaneasa.ro

Abstract: The main concerns and challenges of raising silkworms include economic value, mulberry management, biodiversity conservation of genetic resources, and developing highly productive breeds for genetic variety. This study investigated the relationship between the economic relevance of the products generated throughout the value chain, limitations, and opportunities to generate incomes for sericulture farmers, trends, and perspectives worldwide, particularly in Romania. Seventy-seven publications were considered from online databases. The diversification of products generated at each level of the value chain of silkworm rearing and their multipurpose applications impact social and economic life. Hence, silk is well known as a valuable biomaterial for industry, suitable for textile and medicine. There are several arguments to use silkworms in human food even though they are not yet authorized as edible insects at the European level. Thus, as a nutrient-rich by-product, silkworm pupae (extract, cakes, and oil) have medicinal properties and can be used for human and animal nutrition. Sericin, silk fibroin, and chitin are bioactive compounds in cocoons and pupae with pharmacological implications and drug composition, while biomass is suitable for biodiesel and excreta for compost. The farmers' attitudes and mentality associated with political circumstances influence the perspectives for the sericulture field. Due to the high likelihood of using their products, small-medium-scale farmers might benefit sericulture by identifying new sales marketplaces and finding new beneficiaries for directing their multiple products. The funds allotted by government subventions for supporting this fascinating activity and opportunities for jobs may aid in encouraging to start of a new sericulture business or to contribute developing the existing one.

Keywords: mulberry; silkworm; pupae; silk; economic; biodiversity; conservation

1. Introduction

Climate change, biodiversity as an important driver of food safety and security, population increase, and the identification of novel protein alternatives to those derived from animals are the main issues currently on a worldwide scale.

Insects have been consumed for thousands of years and can be utilized as an alternative source of protein and for waste disposal [1]. For example, insects such as the black soldier fly, which can digest waste materials and generate high-protein larvae, are employed in modern systems [2]. Fisheries may be maintained using insects that can substitute for fishmeal in an aquaculture diet. The use of insects for livestock feeding was described previously by several authors [3–6].

Among insects, the silkworm *Bombyx mori* occupies a special position, being an excellent lepidopteran species representative of numerous scientific investigations [7]. *B. mori* was domesticated and developed by human-driven selection from a wild origin since ancient times. Many years ago, silkworm genetic stocks were conserved in research facilities [8].

The silkworm can be considered from the economic point of view to be one of the most significant insects. It has been providing important benefits to humans, and it has continued developing thanks to its many practical applications, including using it as a model organism for medical purposes [6,9–11].

Presently, UNESCO has designated sericulture as an intangible cultural heritage. In addition to producing several beneficial by-products for human society (pupae, biomass, excreta, mulberry leaves extract, etc.) and adding value through a wide range of applications in economic sectors [12,13], sericulture is required to produce silk fibre for fashionable clothing, shoes, bags, wallpapers, and other fabric items [14].

The economic potential of the sericulture sector, which starts with the production of mulberries (*Morus* sp.) and continues with the rearing of silkworms (upstream) to obtain silk threads (downstream), depends on how this sector can add value and diversify the way of valorisation of the primary product and by-products [15].

It was acknowledged that the production of natural silk threads and the rearing of silkworms, both being historic traditional occupations in the world, would be comparable to a veritable endless "*gold mine*" if all the conditions for doing so would be exploited to the maximum possible extent. In this assumption, experts recognize that increasing and promoting the production of silkworm cocoons are essential to preserve the country's pedoclimatic conditions and the upgrading of this well-earned sector over many decades [16].

Silk production accounts for around 0.24% of all fibre use or about 202,000 metric tonnes annually. *B. mori* provides probably 99% of the silk production [17], although it is challenging to determine the exact global value due to the absence of accurate statistics on completed silk products in the majority of importing nations. China is the leader in production, followed by India, which together account for 95% of global silk manufacturing [18]. On the other hand, although silk thread together with cotton, jute, linen, wool (natural fibres), and nylon and rayon (artificial fibres) is among the most popular known types of fibres, competition among them on the market is obvious, and quality is frequently sacrificed in favour of the price criteria factor when making purchasing decisions.

Silk is a valuable biomaterial that can become more significant in the current textile/fashion value chain framework, although the production of the world's total textiles is relatively low [19]. Under the new market conditions based especially on artificial fibres, silk fibres declined, unable to compete, if we consider the purchasing power of a significant market segment, even though there are several other factors, such as its natural protein fibres. Furthermore, the preservation value of *B. mori* species can support the sustainability of the sericulture sector.

The by-products resulted at each phase along the technological value chain contribute substantially to opening new perspectives for development.

The pupae are a highly nutrient-dense by-product recommended as an edible by-product from the silkworm, which is left after reeling silk. The pupae are utilized in pharmaceutical research after being further treated to extract nutrients and active substances. Silkworm pupae bioactive compounds act for health. The varied medicinal benefits can be remarkable.

The future of sericulture depends on a paradigm shift toward reducing environmental effects and enhancing economic and social life by valuing all products produced by this sector.

This paper investigated the relationship between the economic relevance of the products generated throughout the value chain, limitations, and opportunities to generate incomes for sericulture farmers, trends, and perspectives worldwide, particularly in Romania. We also pointed out the perspective of seeing silkworms as a contributor to decrease *protein hunger* being recommended as suitable alternative protein sources.

2. Materials and Methods

To explore the potential applications of the diverse range of products and by-products generated from sericulture associated with economic and biodiversity implications, we considered publications accessible in Elsevier, MDPI, PubMed, Web of Science, Research Gate, Springer, and Wiley Online Library databases. For easier searching, we used the following key words: "mulberry", "silkworm", "pupae", "silk", "economic trend in sericulture", "biodiversity", "conservation", "compost", "silkworm larvae", "edible insects", "silkworm by-products", "environment", and "silkworm market". The criteria for selecting the publications consisted of the topic of the papers, the English language or bilingual edition with one language being English, and the years. The articles between 2012 and 2023, with few exceptions, were selected attempting to meet the eligibility criteria. The data were collected from over 100 publications, and finally, 83 studies were considered.

3. From the Past to the Present State of Sericulture

The current uncertainty could be clarified with a quick trip into the past. The molecular analyses showed that the silkworm *B. mori* originated from the *B. mandarina* around 4600 years ago in China [20] (quoted by [21]).

The history of silk manufacturing and weaving is reflected in folklore. Undoubtedly, this industry started about 3000 years BC in China. When it was found that approximately 1 km (1000 yards) of thread that makes up a silkworm's cocoon could be reeled out, spun, and woven, the sericulture industry in rural China rapidly gained popularity. According to the Chinese tradition, Lei Zu, the mythical wife of Huangdi, the Yellow Emperor, taught the Chinese people the technique; throughout the history, the empress was symbolically linked to sericulture [14,22,23].

Since approximately 2000 years ago, when the so-called *"Silk Road"* connected Europe to Asia, the silk trade was considered a significant industry preparing the globalization's beginning [19].

Due to industrial development, pollution, and the silkworm disease (pebrine), the sericulture industry has declined since approximately 1873, particularly in Europe. Thus, similar to other species, *B. mori* became endangered, leading to a detrimental effect on biodiversity preservation. The mulberry trees were neglected, and the dead ones have yet to be replaced. The output was restricted to meeting local demands.

Tanase [24] pointed out that the heritage of the silkworm breed depends on mulberry management, maintenance, and utilization of modern techniques necessary to obtain up to 30 tons/ha of leaves.

According to Pau et al. [25], cited by Pop et al. [26], Romania possessed a variety of mulberry species, comprising 49 foreign breeds and hybrids from Japan, China, Russia, Bulgaria, and India in addition to roughly 10 local types.

Over the 20th century, this sector's development experienced significant variations. High levels of hazardous gases, liquid chemicals, and non-biodegradable solid waste were discharged into the nature from household buildings, cars, and industry due to increasing urbanization and the rising human population [27]. Sericulture is one of the industries that was impacted. After this period, silkworm growth developed steadily until the beginning of the 21st century. The specific activities remained at a lower level.

4. Climatic Conditions

No populations of *B. mori* still exist in the wild due to their intense domestication [28], which led to the difficulty of adapting to unfavourable climatic conditions, particularly high temperatures.

According to Rohela et al. [27], mulberry trees are widely distributed across the continents in various environments (from temperate to tropical surfaces), suggesting that they are more tolerant of changing environmental conditions. Hence, mulberry can adapt to different climates, soil types, and altitudes (up to 4000 m above sea level); however, it is resistant only with the ensured light, temperature, water, wind, and other climatic factors [29].

Mulberry trees have a high rate of carbon sequestration that makes it excellent for removing gaseous carbon emissions from the environment [27]. Giacomin et al. [30] and Hăbeanu et al. [6] mentioned that mulberry trees produce leaves for silkworm larvae intake, and more of that can retain carbon. *B. mori* is also very sensitive to pesticides.

One kilogram of raw silk requires approximately 200 kg of mulberry leaves to manufacture it. When the mulberry tree matures in the third year, it can produce around 2 kg of leaves. Mulberry fields per cultivated area reduce CO_2 eq. by about 735 times the weight of synthetic silk fibre. On the contrary, fresh pupae are by-products that pollute the environment and emit an unpleasant odour in the surrounding communities. In the space where silk is produced, large-scale pupae disposal can have detrimental environmental effects [6].

Romania has a mild climate between temperate and continental due to its location in the southeast part of the European continent. The country's diverse relief affects climatic conditions in specific ways. Its geological subregions, which encompass hill, plain, and mountainous areas, are related to climate fluctuation. In the south, the average annual temperature is 11 °C, while in the north, it is 8 °C.

The most important Romanian centres where sericulture is practised to high standards are the Research Station for Sericulture Baneasa, Bucharest, (RSSB) and the Global Centre of Excellence for Advance Research in Sericulture and Promotion of Silk Production (GCEARS-PSP). While the RSSB is situated in the southern lowland region characterized by a humid continental climate with an average annual temperature of 10.8 °C, the GCEARS-PSP is located in the heart of the Transylvania region characterized by a moderate continental environment with oceanic influence and an average annual temperature of 8.2 °C. The conserved genetic resources are the *B. mori* genus.

5. Mulberry Sustainability by Its Applications

Mulberry (genus *Morus*, family *Moraceae*) is a significant economic component in sericulture, the starting point since the quality of the cocoon and the quantity of leaves produced per unit of area are strongly correlated.

The mulberry tree is a perennial plant with significant traits, including higher leaf production, a shorter gestation time, and a greater capacity to adapt to the environment [27,31]. Ninety percent of the world's raw silk output comes from silk from silkworm *B. mori* [28]. This considerably improves many people's lives worldwide. Moreover, it has been stated that the mulberry plant offers a range of nutritional and therapeutic benefits, including its use in livestock feed and well-recognized medical advantages [32].

The mulberry leaves, fruits, and wood push the limits of sustainable development through multiple applications from an economic standpoint. Firstly, mulberry leaves are well known for being the only nourishment source for feeding and raising *B. mori* silkworms, although artificial diets have been adopted [33]. According to Srivastava et al. [34], leaves powder contains key nutrients, such as protein (5–7.3%), fat (2–5%), ash (14.59–17.24%), carbohydrates (9.70–30%), and energy (113–224 kcal/100 g). Cai et al. [35] found condensed tannins, jasmonic acid, anthocyanins, flavonoids, stilbene, and terpenoids, as bioactive compounds required as well for physiological processes in the larva body (growing, reproduction, silk production). One metric tonne of mulberry leaves will yield approximately 25 to 30 kg of cocoons [21].

Mulberry fruits, rich in protein and vitamins, have long been consumed worldwide in human and animal diets. Mulberry products, including juice, tea, wine, and jam, are worthy of inclusion in human diets because of their rich composition in antioxidants, vitamin C, potassium, magnesium, iron, and vitamins K and E.

Several studies have recently investigated the antioxidant potential of extracts from various mulberry plant components, including the leaves, roots, and fruits [36,37]. Furthermore, numerous studies have shown antioxidant, antiviral, anti-inflammatory, hypolipidemic, anti-hyperglycaemic, neuroprotective, anti-HIV, antihypertensive, and cytotoxic properties for mulberry leaves [6,38].

In conventional and contemporary medicine, fruits and leaves are recognized for their antidiabetic properties [21,39]. Traditional Chinese medicine uses mulberry leaf tea with flavonoids and polyphenol antioxidants. The tea has anti-inflammatory properties and reduces cholesterol. Recent studies have also shown mulberry leaf tea has therapeutic benefits for treating diabetes by decreasing blood sugar and inhibiting carbohydrate absorption [40].

Ghosh et al. [31] provided a detailed description of mulberry's economic importance. Thus, the protein-rich mulberry leaves of *Morus alba* can be used successfully as animal feed and can help in lowering blood pressure and cholesterol levels in humans.

Although they were considered waste, the mulberry leaf stalks, and leftovers, such as twigs and shoots, can add value to ruminants and monogastric diets [41–46]. The trunks' wood is valuable for manufacturing paper, handicrafts, cabinetry, handles, and as a fuel source [21].

The main issues that affect the silkworm farmers include mulberry and silkworm rearing management practices, mulberry field area, low quality of silkworm breeds, inadequate infrastructure, a lack of room for raising silkworms, a lack of sufficient start-up equipment for farmers, and low acceptance of new technologies [47].

6. Silkworm Products—Economic Impact

Although there has been a noticeable improvement in silkworm breeding and the creation of new varieties, sericulture development still faces substantial challenges both upstream and downstream. The availability of high-quality eggs, efficiency of farms, health management, a lack or insufficiency of governmental support, unsustainable production, and competition from imported products are the predominant impediments to its development on both the upstream and downstream levels [17]. Sericulture's profitability can be boosted by legislation, investments, and cutting-edge procedures and methods.

Throughout the technological flux of growing silkworms, products (silk thread, pupae cake, pupae oil, larvae and excreta, moths, eggs) are generated at each phase with excellent application in various fields. This increases the incomes of silk farmers and from other related industries.

For instance, the manufacturing of silkworm silk thread (the main product considered a natural protein filament and finest yarns) had a considerable impact on economic development since it formed the basis of rural and urban economy areas of Asia, Europe, and American countries.

According to the FAO report, at the global scale, between 70,000 and 900,000 M.T. of silk is produced yearly, while the demand is rising by 5%. The need for silk will grow due to the population growth and the increased desire for attractive clothing items brought on by rapidly changing fashion trends in highly developed nations.

The high value of silk yarns explains why they have been a crucial part of cross-continental commerce for many generations. Major economic, social, and cultural contacts have resulted from these exchanges over shifting continental and marine routes, but their geopolitical and symbolic long-term significance is still obvious today. The silk production, processing, and trade also serve as the foundation for a tangible cultural heritage by aiding in the development of skills, work organization techniques, encouraging technical and technological transfers, and designing rural and urban landscape types. These traits of sericulture activities make them especially well-suited for addressing economically important topics from a global viewpoint.

Data Bridge Market Research [48] reports that the textile sector has experienced tremendous expansion, leading to the increased demand for silk. Producing silk does not require sophisticated machinery and equipment. These elements support the market's growth. The global silk market is anticipated to grow by USD 32.06 billion by 2029 with a Compound Annual Growth Rate (CAGR) of 8.30% from 2022 to 2029. According to Market Data Forecast [49], the CAGR is predicted to reach 11.32% from 2023 to 2028, while Mordor Intelligence [50] forecasted a CAGR of 5.5% for the 2018–2028 period. The Observatory of Economic Complexity (OEC) [51] reported the total silk transaction amount of USD 1.27 billion in 2020, which contributed to making silk the 96th most traded product in the world. Raw silk exports increased from USD 245 million in 2020 to USD 269 million in 2021.

According to OEC statistics, Romanian silk exports represented 2.07% of the total export value, while imports comprised 13.9%.

Silk fibrous proteins (sericin and fibroin) are beneficial compounds produced in the sericogenic glands of the silkworm through biosynthesis. Sericin, which makes up 25–30% of the protein, surrounds the fibroin fibre with a series of sticky layers that contribute to forming the cocoon. Sericin binds the silk threads around each other, ensuring the cohesiveness of the cocoon [52]. Biodegradable polymers that are favourable to the environment can be produced by mixing sericin with other resins. Sericin and fibroin can be used to create membranes for separation procedures [53].

According to Saric and Scheibel [53], researchers from several fields are looking into using recombinant silk proteins as green bio-polymers. Silk materials may also be macroscopically functionalized to create composite materials that offer remarkable versatility, such as for medical uses in diagnostics, biosensors, or tissue regeneration. Silk is also promising for various clinical and biological applications [54]. Similarly, Kundu et al. [54] highlighted fascinating technical progress made by using bioengineering in implantable optics, photonics, and electronics.

Silkworm pupa characteristics, nutritional value, and applications in different fields were detailed in our previous paper [6]. Thus, we focused on the excellent nutritional characteristics emphasizing fatty acids beneficial to health (omega-3 particularly) and wide applicability in human food and animal feed, medicine, and environmental impact reduction. The varied and intricate medicinal benefits, such as immunomodulatory, antibacterial, anticancer, antioxidant, hepatoprotective, antifatigue, and antiapoptotic actions, are also remarkable. Silkworm pupa bioactive compounds act by reducing plasma triglycerides. These bio-compounds act to treat wounds and control blood sugar levels, lower blood pressure, prevent thrombosis and arteriosclerosis, boost cell vitality, and enhance the body's defences in metabolic disorders [6,55,56].

Among the benefits of silkworm proteins and hydrolysed peptides are increased immunity and anticancer and antioxidant properties [57].

From a dietary perspective, the larvae and pupae are edible parts of the silkworms and are eaten by the inhabitants of China, Japan, Korea, Thailand, and Northeast India [58]. Due to their high protein content, silkworm pupae have been sold for centuries in various marketplaces and are regarded as an economically viable commodity [15].

Sericulture focuses on many methods that might boost the farms' incomes by manufacturing cocoons. Small-scale farmers need to identify new revenue streams, including the sale of cocoons and other applications of by-products. Silkworm rearing allows for a cheap supply of biomass, which is potentially a beneficial source for biogas generation [59]. In a study in 2019, Łochyńska and Frankowski [60] presented an option with encouraging results of fertilizer organic hemp culture (*Cannabis sativa* L.) with larva excrements.

Traditional Asian medicine has employed excreta therapeutically to treat various diseases [59]. Since it contains a lot of flavonoids, chlorophyll, alkaloids, carotenoids, and lutein, silkworm excreta may have considerable antioxidant action [6].

The litter of silkworms (leftover leaves and faeces) can be used as organic manure [61].

Over time, due to concerns about animal welfare, scientists have been obliged to restrict the number of vertebrates used as experimental animal models. Consequently, other animal models that do not need ethics committee approval have been considered. A richness of genetic resources, an evident genetic pedigree, and a short generation period with a 25–30-day larval stage are some traits that make using silkworms a more favourable, simpler, and less expensive model for studies [10]. Kodama et al. [62], quoted by Meng et al. [10], developed the first silkworm larva viral infection model for medication therapy. The findings of injecting the virus into the silkworm's haemolymph showed that nalidixic acid might suppress the growth of the flacherie virus and nuclear polyhedrosis virus (BmNPV) and shield silkworms against infection with related viruses.

Due to their great susceptibility to pesticides, antibiotics, pathogenic fungi, and germs that are harmful to humans, silkworms have become a paradigm for natural immune activation, drug screening and kinetic testing [10]. In addition, research on the pathophysiology and human microorganism toxicity has rapidly improved in recent decades. It was discovered that the chitosan isolated from silkworm pupae improved economic parameters. Hence, efficient pupa reutilization might be employed to produce a high-potential raw material for numerous biomedical processes by turning it into beneficial bioproducts [63].

In contemporary research, the silkworm is used as a model organism to investigate human tumours, degenerative diseases, and metabolic diseases.

Epigenetics investigates genetic variation at the chromatin and DNA alteration levels without variations in gene sequence [64]. In cutting-edge research and technology, it is a crucial information carrier. Sericulture research advanced in 2010 by creating the first silkworm methylation map (as an epigenetics marker).

Aznar-Cervantes et al. [65] showed that it was possible to use the silkworm to assess the hypoglycaemic activity of various products originating from sericulture by dietary addition after promoting glucose or sucrose-induced hyperglycaemia.

Future directions and targets may be established and investigated as science continuously develops.

7. Edible Silkworm—Mentality and Perspectives

7.1. Silkworm as a Source of Food

A 60% increase in the requirement for food is anticipated by 2050 [66,67]. Since the Novel Food Regulation was enacted on 1 January 2018, EFSA [68] has received many applications covering various novel and traditional food sources. They include foods made from algae, non-indigenous fruits, herbal compounds derived from plants, and different edible insect species. Silkworms *B. mori* have been considered as a suitable insect for human consumption as a substitute for animal protein, but did not yet receive official authorization as edible insects.

Edible insects are a distinct food ingredient that continues to gain value in improving global food security and providing an attractive alternative to meat. Insects can contribute to food production and reduce the adverse effects of climate change [69]. More than 2000 edible insect species, including silkworms, are consumed globally [67] in raw or processed form [69].

A comprehensive nutritional comparison between *B. mori* and meat from different species was provided by Orkusz et al. [67]. Accordingly, the energy value (kcal/100 g) of edible parts of *B. mori* ranges between 171.27 and 229, for horse meat it is 109, for pork shoulder 13.2, for beef shoulder 112, for rabbit carcass 123.24, for goose carcass 140.63, for duck carcass 199.04, for turkey breast 83, for chicken breast 98, and for chicken drumstick 125. Regarding protein level (g/100 g), the higher value was reported for *B. mori* (17.9–23.1), followed by chicken breast and horse meat (21.5%), while the lower value was found in the goose carcass. For the poor rural population especially, *B. mori* represents a source more attractive from the nutritional point of view than meat. The advantages include fast growth, more efficient feed conversion, and minimum resource requirement.

Although insects present many advantages, most consumers are wary of their safety in most developed nations and reluctant to include insects in their diets. However, the studies concerning edible insects are limited.

In Romania, similar to many Western countries, there is a strong mentality against consuming edible insects, despite the EU opinion and advice due to respect for culinary traditions, mentality, and lack of sufficient information. Zugravu et al. [70] pointed out the importance of knowledge as a determinant factor for accepting the intake of edible insects. The study of Zugravu et al. [70], including Romania, as part of a global investigation, used a questionnaire translated into Romanian to assess attitudes against eating insects and find modifiable factors associated with reluctance or aversion. Even though most people do not find insects appetizing, those who know more about them may be willing to try these novel food products because they relate them to protein or other nutrients content. Nonetheless, overcoming aversion will be crucial in accepting insect ingestion, as eating patterns change hesitantly and gradually.

As an argument, the FAO [71], cited by Zugravu et al. [70], reported that a significant portion of the population will not have enough water by 2025. Raising and processing insects requires less water. Certain insects, such as the yellow and smaller mealworms and silkworms, can withstand drought.

According to Wu et al. [57], 26 proteins from silkworm pupae have been identified as allergens. Phytate, phytic phosphorus, tannic acid, alkaloid, flavonoids, saponin, and oxalate are some of the antinutrients in silkworm pupae that have drawn the attention of several authors; however, their levels do not present a risk to humans [6,72,73].

The WHO and International Union of Immunological Societies (WHO/IUIS) Allergen Nomenclature Sub-Committee (www.allergen.org) [74] have not officially confirmed or registered any allergens of silkworm pupae aside from this arginine kinase (formally known as Bomb m 1). Therefore, a thorough risk evaluation of possible allergic risks is necessary before establishing silkworm pupae as a food source. Few studies evaluate silkworm pupa allergens, and most tend to base them on case reports rather than regularly using standardized tests. Although 26 silkworm pupa proteins have been identified as allergens, little is known about their immunological characteristics [57].

7.2. *Silkworm, a Valuable Feedstuff for Animal Feeding*

There is an increased interest in using silkworms in animal feeding due to their protein levels and protein digestibility (76–98%), lipids, and essential fatty acids [6], which emphasize their significance in the current trends in feed science and related sectors. The effects of using silkworm pupae in ruminants and monogastric species, including chickens, pigs, rabbits, and fish, were studied by many authors [75–78]. Silkworm pupae are the potential to replace soybean meal or fish meal in animal feeding. Silkworm pupa meal is a valuable and less expensive alternative protein source having a little lower quality than fish meal. The properties of hen eggs and yolk colour improved when silkworm pupae were used as a protein supplement [5]. In monogastric diets, silkworm pupae did not significantly impact growth and feed efficiency parameters.

The results demonstrated that silkworm pupae can stimulate rainbow trout's immune system and cause some anaemia-related symptoms.

There is little information available about using silkworm pupae in pig diets. However, the pupal powder is advised for pig feeding due to its high protein content [79], albeit the increased oil content may cause an adverse reaction. Interestingly, the protein in pupae is superior to that in fish and soybean meal [5].

8. Employment Opportunity

Sericulture is a low-start-up-cost, employment-focused industry that generates income. The agro-based economic development is depicted as bridging the gap between market-oriented development and various development policy issues and goals, particularly poverty reduction, climate change adaptation, and ecological and economic transformation.

Every strategy, action, and link in the value chain must consider the broader structure having as central point humans and their needs. The sericulture industry would significantly impact the modern economies of many nations worldwide.

Commercial silk offers a high return to the producers who live in rural areas of developing nations, and the silk industry requires only modest capital. Thanks to the sericulture sector, millions of people from rural areas are provided with jobs, which prevents and reduces migration to urban areas. More than this, the development of the silk trade was aided by the expansion of the global market, which ensured significant financial inflows into the payment systems of developing nations [80].

Each stage of the technological process involved in sericulture provides products and by-products with a wide range of applications that have been identified and become accessible over time. The question raised is how farmers who use sericulture wastes might benefit.

According to a statement established by the EU in October 2007, applicable in Romania, sericulture producers can obtain funding from the EU of about 134 euros for each carton of silkworm eggs if they deliver the products to a licensed processing centre.

To be eligible for the financial support, breeders must purchase boxes containing a minimum of 20,000 eggs from licensed distributors (e.g., in Romania, Research Station for Sericulture Baneasa, Bucharest), then procure and distribute a minimum of 15–20 kg of silkworm cocoons per box to an authorized processing centre. Breeders must also be members of a sericulture association.

9. Sericulture Development Policy—Perspectives

The practice of eating insects is known as entomophagy, and historical evidence supports this assertion [81].

For a higher economic impact, some more productive mulberry varieties, conservation of silkworm genetic resources, obtaining a new silkworm breed resistant to harmful climatic conditions and diseases, and biodiversity conservation are major challenges. At many national and international forums, the concerns of the access to genetic resources, their sustainable use, benefit sharing, and the farmers' rights are being debated.

At present, few farmers practice sericulture, and even significant subsidies may be obtained. The current government policy is directed to encourage and support the start-up of businesses, particularly in rural areas.

Romania's decision-making factors involved adopting measures to revive the sector, implying specialists for know-how transfer.

Moise et al. [82] and Dezmirean et al. [83] emphasized the significance of the Romanian GCEARS-PSP Centre, which has been recognized by the International Sericulture Commission since 2014, and whose main objectives include developing cutting-edge sericulture research and analyses by using modern methods, building a gene pool for various silkworm breeds, and trying to promote silk output. The revitalization of the Romanian sericulture industry is the primary goal of GCEARS-PSP. A framework for restarting the activity at the Research Station for Sericulture Baneasa, Bucharest, (RSSB) was also established in 2022 due to a government directive using updated coordinates.

Given the importance of raising silkworms via their diversity of products along the value chain and their wide range of possible applications with economic relevance, the involvement of policies, and the expertise of specialists in the sector's relaunch, the projection may be seen as a positive one (Figure 1). A key component of future development is offering technical support for research activities.

Given the financial assistance received by those engaged (specialists, farmers, and processors), silkworm breeding in Romania has great potential. However, it is crucial to keep in mind that the magnitude of subsidies and the development of economic variables are correlated, but it is necessary to keep in mind that the extent of subsidies and the development of financial elements are related.

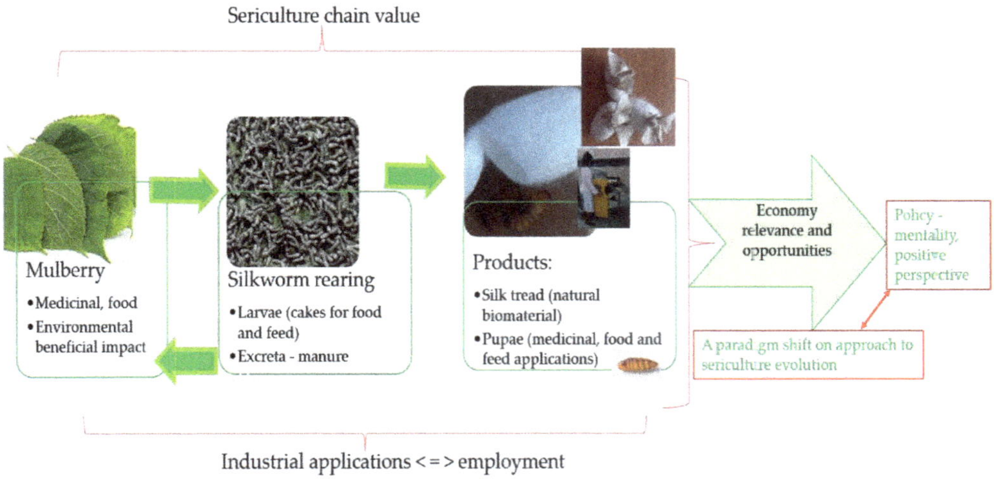

Figure 1. Economic opportunities and perspective of sericulture.

10. Conclusions and Future Directions

A paradigm shift on the approach to sericulture evolution is required. Although sericulture has seen a downturn, the paradox consists in the fact that its economic viability is strongly maintained by a number of factors. The economic potential of sericulture is determined by the ability to value its main output (silk thread) and by-products (pupae, excreta, mulberry leaves, etc.) in various ways (medicine, model insects in genetic and physiology research, food for humans and animal feeding, as compost, manure, biodiesel, etc.). The profitability depends firstly on the mulberry tree production, area, and preservation, and maintaining the habitat despite industrialization. Silkworms are an important contributor to biodiversity and become more and more important as a key factor for food safety and security.

Debatable economic topics include biodiversity conservation, searching for new markets for silk, sharing expertise, know-how transfer, adaptability and shifts in public perception, and regulations at both national and international levels. These issues may be overcome by considering the substantial economic potential of using different silkworm products, the opportunity for sericulture farmers to enhance their earnings, and the promotion and encouragement provided by considerable subsidies.

To increase the value of sericulture, it is also essential to preserve the habitat of the mulberry tree, which is the only feed supply for the silkworm *B. mori*.

Another crucial step is to market and find new applications for by-products such as pupae, excreta, and mulberry leaves, silk threads (as principal commodities).

Providing technical support for research activities is an essential point in future development.

From an economic point of view, another way to boost sericulture is to encourage people from rural areas to start or expand their silkworm-rearing businesses by supplying them with substantial subsidies.

As supplementary arguments, silkworm pupae, besides multiple effects on the medical field and in animal feeding, can provide an alternative to other animal proteins (such as meat, for example).

Author Contributions: Conceptualization, M.H.; methodology, M.H. and A.G.; software, M.H.; validation, T.M.; formal analysis, A.G.; investigation, M.H.; resources, T.M.; data curation, M.H. and A.G.; writing—original draft preparation, M.H.; writing—review and editing, M.H. and A.G.; visualization, A.G.; supervision, T.M.; funding acquisition, A.G. All authors have read and agreed to the published version of the manuscript.

Funding: The APC was funded by the Romanian Ministry of Agriculture and Rural Development.

Institutional Review Board Statement: Not applicable.

Data Availability Statement: Not applicable.

Conflicts of Interest: The authors declare no conflict of interest.

References

1. Govorushko, S. Global status of insects as food and feed source: A review. *Trends Food Sci. Technol.* **2019**, *91*, 436–445. [CrossRef]
2. Bessa, L.W.; Pieterse, P.; Marais, J.; Dhanani, K.; Hoffman, L.C. Food safety of consuming black soldier fly (*Hermetia illucens*) larvae: Microbial, heavy metal and cross-reactive allergen risks. *Foods* **2021**, *10*, 1934. [CrossRef] [PubMed]
3. Khatun, R.; Azmal, S.A.; Sarker, M.S.K.; Rashid, M.A.; Hussain, M.A.; Miah, M.Y. Effect of silkworm pupae on the growth and egg production performance of Rhode Island Red (RIR) pure line. *Int. J. Poult. Sci.* **2005**, *4*, 718–720.
4. Ullah, R.; Khan, S.; Khan, N.A.; Mobashar, M.; Lohakare, J. Replacement of soybean meal with silkworm meal in the diets of White Leghorn layers and effects on performance, apparent total tract digestibility, blood profile and egg quality. *J. Vet. Sci. Res.* **2017**, *5*, 200–207.
5. Priyadharshini, P.; Joncy, M.A.; Saratha, M. Industrial utilization of silkworm pupae—A review. *Int. Acad. Res. Multidiscip.* **2017**, *5*, 62–70.
6. Hăbeanu, M.; Gheorghe, A.; Mihalcea, T. Nutritional value of silkworm pupae (*Bombyx mori*) with emphases on fatty acids profile and their potential applications for humans and animals. *Insects* **2023**, *14*, 254. [CrossRef]
7. Panthee, S.; Paudel, A.; Hamamoto, H.; Sekimizu, K. Advantages of the silkworm as an animal model for developing novel antimicrobial agents. *Front. Microbiol.* **2017**, *8*, 373. [CrossRef]
8. Cappellozza, S.; Casartelli, M.; Sandrelli, F.; Saviane, A.; Tettaman, G. Silkworm and Silk Traditional and Innovative Applications. *Insects* **2023**, *13*, 1016. [CrossRef]
9. Nwibo, N.N.; Hamamoto, H.; Matsumoto, Y.; Kaito, C.; Sekimizu, K. Current use of silkworm larvae (*Bombyx mori*) as an animal model in pharmaco-medical research. *Drug Discov. Ther.* **2015**, *9*, 133–135. [CrossRef]
10. Meng, X.; Zhu, F.; Che, K. Silkworm: A promising model organism in life science. *J. Insect Sci.* **2017**, *17*, 97. [CrossRef]
11. Neelaboina, B.K.; Shivkumar, M.N.A.; Ghosh, M.K. Studies on the performance of some silkworm, *Bombyx mori* L., breeds in temperate region of Jammu and Kashmir, India. *Int. J. Curr. Microbiol. Appl. Sci.* **2018**, *7*, 2192–2201. [CrossRef]
12. Jasmine, S.; Mandal, B.B. Chapter 10. Types and Properties of Non-Mulberry Silk Biomaterials for Tissue Engineering Applications. In *Silk Biomaterials for Tissue Engineering and Regenerative*; Kundu, S.C., Ed.; Woodhead Publishing: Cambridge, UK, 2014; pp. 275–298. [CrossRef]
13. Oduor, E.O.; Ciera, L.; Adolkar, V.; Pido, O. Physical characterization of eri silk fibers produced in Kenya. *J. Nat. Fibers* **2019**, *18*, 59–70. [CrossRef]
14. Gjurašić, M.; Đurović, T. Development of sericulture in the eastern Adriatic during the Austrian administration. *Athens J. Hist.* **2023**, *9*, 9–52. [CrossRef]
15. Sharma, V.; Rattan, M.; Chauhan, S.K. Potential use of sericultural by products: A review. *Pharma Innov.* **2022**, *SP-11*, 1154–1158.
16. Tzenov, P.; Cappellozza, S.; Saviane, A. Black, Caspian Seas and Central Asia Silk Association (BACSA) for the future of seri-culture in Europe and Central Asia. *Insects* **2022**, *13*, 44. [CrossRef]
17. Andadari, L.; Yuniati, D.; Supriyanto, B.; Murniati; Suharti, S.; Widarti, A.; Steven, E.; Sadapotto, A.; Winarno, B.; Minarningsih; et al. Lens on tropical sericulture development in Indonesia: Recent status and future directions for industry and social forestry. *Insects* **2022**, *13*, 913. [CrossRef]
18. Padaki, N.V.; Das, B.; Basu, A. *Advances in Understanding the Properties of Silk. Advances in Silk Science and Technology*; Basu, A., Ed.; Woodhead Publishing Series in Textiles; Woodhead Publishing: Cambridge, UK, 2015; pp. 3–16. [CrossRef]
19. Giora, G.; Marchetti, G.; Cappellozza, S.; Assirelli, A.; Saviane, A.; Sartori, L.; Marinello, F. Bibliometric analysis of trends in mulberry and silkworm research on the production of silk and its by-Products. *Insects* **2022**, *13*, 568. [CrossRef]
20. Arunkumar, K.P.; Metta, M.; Nagaraju, J. Molecular phylogeny of silkmoths reveals the origin of domesticated silkmoth, *Bombyx mori* from Chinese *Bombyx mandarina* and paternal inheritance of *Antheraea proylei* mitochondrial DNA. *Mol. Phylogenet. Evol.* **2006**, *40*, 419–427. [CrossRef]
21. Altman, G.H.; Farrell, B.D. Sericulture as a sustainable agroindustry. *Clean. Circ. Bioecon.* **2022**, *2*, 100011. [CrossRef]
22. Gaston, G.O. History of sericiculture in France. *Eur. J. Res. Soc. Sci.* **2017**, *5*, 4–58.
23. Łochyńska, M. History of sericulture in Poland. *J. Nat. Fibers* **2010**, *7*, 334–337. [CrossRef]
24. Tanase, D. The agro productive characterization of the mulberry varieties used in the amelioration programs. *Sci. Pap. Anim. Sci. Biotechnol.* **2007**, *40*, 141–149.
25. Pau, E.; Constantinescu, M. Solutions for sericulture reorganization in Romania. In Proceedings of the First International Conference "Sericulture—From Tradition to Modern Biotechnology", Cluj-Napoca, Romania, 17–18 April 2008; pp. 133–140.
26. Pop, L.L.; Mărghitaș, A.L.; Dezmirean, D.; Bobis, O.; Moise, A.; Pasca, C. Sericiculture industry in Romania—Analysis on current situation and prospects of development. *Sci. Pap. Ser. D Anim. Sci.* **2018**, *LXI*, 251–258.
27. Rohela, G.K.; Shukla, P.; Muttanna; Kumar, R.; Chowdhury, S.R. Mulberry (*Morus* spp.): An ideal plant for sustainable development. *Trees For. People* **2020**, *2*, 100011. [CrossRef]

28. Ruiz, X.; Almanza, M. Implications of genetic diversity in the improvement of silkworm *Bombyx mori* L. *Chil. J. Agric. Res.* **2018**, *78*, 569–579. [CrossRef]
29. Memete, A.R.; Timar, A.V.; Vuscan, A.N.; Miere (Groza), F.; Venter, A.C.; Vicas, S.I. Phytochemical composition of different botanical parts of *Morus* species, health benefits and application in food industry. *Plants* **2022**, *11*, 152. [CrossRef]
30. Giacomin, A.M.; Garcia, J.B., Jr.; Zonatti, W.F.; Silva-Santos, M.C.; Laktim, M.C.; Baruque-Ramos, J. Silk industry and carbon footprint mitigation. *Procedia Eng. Conf. Ser. Mater. Sci. Eng.* **2017**, *254*, 192008. [CrossRef]
31. Ghosh, A.; Gangopadhyay, D.; Chowdhur, T. Economical and environmental importance of mulberry: A Review. *Int. J. Plant Environ.* **2017**, *3*, 51–58. [CrossRef]
32. Sujathamma, P.; Savithri, G.; Kavyasudha, K. Value addition of mulberry (*Morus* sp.). *Int. J. Emerg. Technol. Comput. Appl. Sci.* **2013**, *7*, 352–356.
33. Tassoni, L.; Cappellozza, S.; Dalle Zotte, A.; Belluco, S.; Antonelli, P.; Marzoli, F.; Saviane, A. Nutritional composition of *Bombyx mori* pupae: A systematic review. *Insects* **2022**, *13*, 644. [CrossRef]
34. Srivastava, S.; Kapoor, R.; Thathola, A.; Srivastava, R.P. Nutritional quality of leaves of some genotypes of mulberry (*Morus alba*). *Int. J. Food Sci. Nutr.* **2006**, *57*, 305–313. [CrossRef] [PubMed]
35. Cai, M.; Mu, L.; Wang, Z.; Liu, J.; Liu, T.; Wanapat, M.; Huang, B. Assessment of mulberry leaf as a potential feed supplement for animal feeding in PR China. *Asian-Aust. J. Anim. Sci.* **2019**, *32*, 1145. [CrossRef] [PubMed]
36. Andallu, B.; Varadacharyulu, N.C. Antioxidant role of mulberry (*Morus indica* L. cv. Anantha) leaves in streptozotocin-diabetic rats. *Clin. Chim. Acta* **2003**, *338*, 3–10. [CrossRef] [PubMed]
37. Arfan, M.; Khan, R.; Rybarczyk, A.; Amarowicz, R. Antioxidant activity of mulberry fruit extract. *Int. J. Mol. Sci.* **2012**, *13*, 2472–2480. [CrossRef]
38. Pan, G.; Lou, C.F. Isolation of an 1-aminocyclopropane-1-carboxylate oxidase gene from mulberry (*Morus alba* L.) and analysis of the function of this gene in plant development and stresses response. *J. Plant Physiol.* **2008**, *165*, 1204–1213. [CrossRef]
39. Yuan, Q.; Zhao, L. The Mulberry (*Morus alba* L.) Fruit: A review of characteristic components and health benefits. *J. Agric. Food Chem.* **2017**, *65*, 10383–10394. [CrossRef]
40. Józefczuk, J.; Malikowska, K.; Glapa, A.; Stawińska-Witoszyńska, B.; Nowak, J.K.; Bajerska, J.; Lisowska, A.; Walkowiak, J. Mulberry leaf extract decreases digestion and absorption of starch in healthy subjects—A randomized, placebo-controlled, crossover study. *Adv. Med. Sci.* **2017**, *62*, 302–306. [CrossRef]
41. Olteanu, M.; Panaite, T.; Ciurescu, G.; Criste, R.D. Effect of dietary mulberry leaves on performance parameters and nutrient digestibility of laying hens. *Indian J. Anim. Sci.* **2012**, *82*, 914–917.
42. Olteanu, M.; Criste, R.D.; Cornescu, G.M.; Ropota, M.; Panaite, T.D.; Varzaru, I. Effect of dietary mulberry (*Morus alba*) leaves on performance parameters and quality of breast meat of broilers. *Indian J. Anim. Sci.* **2015**, *85*, 291–295.
43. Ustundag, A.O.; Ozdogan, M. Usage possibilities of mulberry leaves in poultry nutrition. *Sci. Pap. Ser. D Anim. Sci.* **2015**, *LVIII*, 18.
44. Wang, C.; Yang, F.; Wang, Q.; Zhou, X.; Xie, M.; Kang, P.; Wang, Y.; Peng, X. Nutritive value of mulberry leaf meal and its effect on the performance of 35-70-day-old geese. *J. Poult. Sci.* **2017**, *54*, 41–46. [CrossRef] [PubMed]
45. Liu, Y.; Li, Y.; Peng, Y.; He, J.; Xiao, D.; Chen, C.; Li, F.; Huang, R.; Yin, Y. Dietary mulberry leaf powder affects growth performance, carcass traits and meat quality in finishing pigs. *J. Anim. Physiol. Anim. Nutr.* **2019**, *103*, 1934–1945. [CrossRef] [PubMed]
46. Şengul, A.Y.; Şengul, T.; Celik, Ş.; Şengül, G.; Daş, A.; İnci, H.; Bengu, A.Ş. The effect of dried white mulberry (*Morus alba*) Pulp supplementation in diets of laying quail. *Rev. MVZ Cordoba* **2021**, *26*, e1940. [CrossRef]
47. Ssemugenze, B.; Esimu, J.; Nagasha, J.; Masiga, C.W. Sericulture: Agro-based industry for sustainable socio-economic development: A review. *Int. J. Sci. Res. Publ.* **2021**, *11*, 474–482. [CrossRef]
48. Data Bridge Market Research. Available online: https://www.databridgemarketresearch.com/reports/global-silk-market (accessed on 30 March 2023).
49. Market Data Forecast. Available online: https://www.marketdataforecast.com/market-reports/silk-market (accessed on 30 March 2023).
50. Mordor Intelligence. Available online: https://www.mordorintelligence.com/industry-reports/silk-yarn-market (accessed on 30 March 2023).
51. The Observatory of Economic Complexity (OEC). Available online: https://oec.world/en/profile/hs/silk (accessed on 30 March 2023).
52. Zhang, Y.-Q. Applications of natural silk protein sericin in biomaterials. *Biotechnol. Adv.* **2002**, *20*, 91–100. [CrossRef]
53. Saric, M.; Scheibel, T. Engineering of silk proteins for materials applications. *Curr. Opin. Biotechnol.* **2019**, *60*, 213–220. [CrossRef]
54. Kundu, B.; Kurland, N.E.; Bano, S.; Patra, C.; Engel, F.B.; Yadavalli, V.K.; Kundu, S.C. Silk proteins for biomedical applications: Bioengineering perspectives. *Prog. Polym. Sci.* **2014**, *39*, 251–267. [CrossRef]
55. Kim, Y.J.; Lee, K.P.; Lee, D.Y.; Kim, Y.T.; Baek, S.; Yoon, M.S. Inhibitory effect of modified silkworm pupae oil in PDGF-BB-induced proliferation and migration of vascular smooth muscle cells. *Food Sci. Biotechnol.* **2020**, *29*, 1091–1099. [CrossRef]
56. Zhou, Y.; Zhou, S.; Duan, H.; Wang, J.; Yan, W. Silkworm pupae: A functional food with health benefits for humans. *Foods* **2022**, *11*, 1594. [CrossRef]
57. Wu, X.; He, K.; Cirkovic Velickovic, T.; Liu, Z. Nutritional, functional, and allergenic properties of silkworm pupae. *Food Sci. Nutr.* **2021**, *9*, 4655–4665. [CrossRef]

58. Sharma, A.; Kumar Gupta, R.; Sharma, P.; Duwa, A.K.; Bandral, R.S.; Bali, K. Silkworm as an edible insect: A review. *Pharma Innov.* **2022**, *SP-11*, 1667–1674.
59. Łochyńska, M.; Frankowski, J. The biogas production potential from silkworm waste. *Waste Manag.* **2018**, *79*, 564–570. [CrossRef] [PubMed]
60. Łochyńska, M.E.; Frankowski, J. Impact of silkworm excrement organic fertilizer on hemp biomass yield and composition. *J. Ecol. Eng.* **2019**, *20*, 63–71. [CrossRef]
61. Shanmugam, R.; Mohanraj, P.; Krishnamoorthy, S.V.; Chozhan, K. Stimulus of silkworm excreta on quality and quantity of cocoon production. *Int. J. Creat. Res. Thoughts* **2020**, *8*, 1166–1171.
62. Kodama, R.; Nakasuji, Y. Bacteria isolated from silkworm larvae: X. Inhibition of development of viral diseases in gnotobiotic silkworm by nalidixic acid. *J. Insect Biotechnol. Sericol.* **1972**, *41*, 7–41.
63. Pachiappan, P.; Prabhu, S.; Mahalingam, C.A.; Thangamalar, A.; Umapathy, G. In vivo antibacterial effect of chitosan against *Staphylococcus aureus* and *Bacillus thuringiensis* and its impact on economic parameters of silkworm, *Bombyx mori.* L. *J. Pharmacogn. Phytochem.* **2018**, *7*, 2448–2451.
64. Hafner, S.; Lund, A.H. Great expectations e epigenetics and the meandering path from bench to bedside. *Biomed. J.* **2016**, *19*, 166–176. [CrossRef]
65. Aznar-Cervantes, S.D.; Monteagudo Santesteban, B.; Cenis, J.L. Products of sericulture and their hypoglycemic action evaluated by using the silkworm, *Bombyx mori* (Lepidoptera: Bombycidae), as a model. *Insects* **2021**, *12*, 1059. [CrossRef]
66. Food and Agriculture Organization of the United Nation (FAO). *Edible Insects: Future Prospects for Food and Feed Security*; Food and Agriculture Organization of the United Nation (FAO): Rome, Italy, 2013.
67. Orkusz, A. Edible insects versus meat—Nutritional comparison: Knowledge of their composition is the key to good health. *Nutrients* **2021**, *13*, 1207. [CrossRef]
68. European Food Safety Authority (EFSA). Scientific opinion on a risk profile related to production and consumption of in-sects as food and feed. *EFSA J.* **2015**, *13*, 4257. [CrossRef]
69. Imathiu, S. Benefits and food safety concerns associated with consumption of edible insects. *NFS J.* **2020**, *18*, 1–11. [CrossRef]
70. Zugravu, C.; Tarcea, M.; Nedelescu, M.; Nuță, D.; Guiné, R.P.F.; Constantin, C. Knowledge: A factor for acceptance of insects as food. *Sustainability* **2023**, *15*, 4820. [CrossRef]
71. Food and Agriculture Organization of the United Nation (FAO). *Looking at Edible Insects from a Food Safety Perspective. Challenges and Opportunities for the Sector*; Food and Agriculture Organization of the United Nation (FAO): Rome, Italy, 2021.
72. Ji, K.M.; Zhan, Z.K.; Chen, J.J.; Liu, Z.G. Anaphylactic shock caused by silkworm pupa consumption in China. *Allergy* **2008**, *63*, 1407–1408. [CrossRef] [PubMed]
73. Feng, Y.; Chen, X.M.; Zhao, M.; He, Z.; Sun, L.; Wang, C.Y.; Ding, D.F. Edible insects in China: Utilization and prospects. *Insect Sci.* **2018**, *25*, 184–198. [CrossRef] [PubMed]
74. The World Health Organization and International Union of Immunological Societies (WHO/IUIS). Allergen Nomenclature Sub-Committee. Available online: www.allergen.org (accessed on 30 March 2023).
75. Sheikh, I.U.; Banday, M.T.; Baba, I.A.; Adil, S.; Shaista, S.N.; Bushra, Z.; Bulbul, K.H. Utilization of silkworm pupae meal as an alternative source of protein in the diet of livestock and poultry: A review. *J. Entomol. Zool. Stud.* **2018**, *6*, 1010–1016.
76. Asimi, O.A.; Bhat, T.H.; Nasir, H.; Irfan, K. Alternative Source of Protein "Silkworm Pupae" (*Bombyx mori*) in Coldwater Aquaculture. *Int. J. Poult. Fish. Sci.* **2017**, *1*, 1–4. [CrossRef]
77. Herman, R.A.; Yan, C.-H.; Wang, J.-Z.; Xun, X.M.; Wu, C.-K.; Li, Z.N.; Ayepa, E.; You, S.; Gong, L.C.; Wang, J. Insight into the silkworm pupae: Modification technologies and functionality of the protein and lipids. *Trends Food Sci. Technol.* **2022**, *129*, 408–420. [CrossRef]
78. Shakoori, M.; Gholipour, M.; Naseri, S. Effect of replacing dietary fish meal with silkworm (*Bombyx mori*) pupae on hematological parameters of rainbow trout *Oncorhynchus mykiss*. *Comp. Clin. Pathol.* **2013**, *24*, 139–143. [CrossRef]
79. Trivedy, K.; Kumar, S.N.; Mondal, M.; Bhat, C.A.K. Protein banding pattern and major amino acids components in de-oiled pupal powder of silkworm, *Bombyx mori* L. *J. Entomol.* **2008**, *5*, 10–16. [CrossRef]
80. Popescu, A. Considerations upon the trends in the world silk trade. *Sci. Pap. Ser. Manag. Econ. Eng. Agric. Rural. Dev.* **2018**, *18*, 385–400.
81. Liceaga, A.M. Chapter Four—Edible insects, a valuable protein source from ancient to modern times. *Adv. Food Nutr.* **2022**, *101*, 129–152. [CrossRef]
82. Moise, A.R.; Marghitas, L.A.; Bobis, O.; Copaciu, F.M.; Dezmirean, D.S. *Morus* spp. Material conservation and characterization and its importance for Romanian sericulture and GCEARS-PSP development—A Review. *Bull. Univ. Agric. Sci. Vet. Med. Cluj-Napoca Anim. Sci. Biotechnol.* **2018**, *75*, 57–63. [CrossRef] [PubMed]
83. Dezmirean, D.; Mărghitaş, L.A.; Bobiş, O.; Urcan, A.C.; Dezmirean, H.; Paşca, C.; Moise, A.R. Multidirectional activities for gene pool conservation in GCEARS-PSP. *Bull. Univ. Agric. Sci. Vet. Med. Cluj-Napoca Anim. Sci. Biotechnol.* **2018**, *75*, 5–10. [CrossRef] [PubMed]

agriculture

MDPI

Article

Prickly Pear Cladodes as an Alternative Source of Water in Small- and Medium-Scale Yellow Mealworm Rearing

Simona Errico *, Paola Sangiorgio, Stefania Moliterni, Alessandra Verardi, Anna Spagnoletta, Salvatore Dimatteo and Ferdinando Baldacchino

Laboratory of Bioproducts and Bioprocess, ENEA—Trisaia Research Centre, S.S. Jonica 106, Km 419+500, I-75026 Rotondella, Matera, Italy; paola.sangiorgio@enea.it (P.S.); stefania.moliterni@enea.it (S.M.); alessandra.verardi@enea.it (A.V.); anna.spagnoletta@enea.it (A.S.); salvatore.dimatteo@enea.it (S.D.); ferdinando.baldacchino@enea.it (F.B.)

* Correspondence: simona.errico@enea.it

Abstract: New solutions are compulsorily needed to reconcile the enormous and ever-growing request for protein for human nutrition and, at the same time, reduce conventional meat production. This epochal challenge can find a valuable aid to a winning solution in insect rearing. The use of insects as feed and food, far from being a definitive solution to global food shortages, can offer new protein sources and perfectly fit circular economy precepts, yet more so when insects feed on by-products from the agri-food industry. In this scenario, *Tenebrio molitor* (TM) is a concrete alternative. Therefore, making its rearing more sustainable is a prime objective. In this paper, we evaluated the possibility of replacing usual plant sources of wet supplementation used in TM rearing with sustainable alternatives, including the cladodes of prickly pear (*Opuntias ficus indica*, OFI), to reduce the frequency of administration, thus minimizing related labor costs. The alternatives were tested for water content, dehydration, and shelf life to select the best-performing ones. On the selected matrices, we evaluated the preference of the larvae and their palatability because a matrix may be convenient and sustainable but not appreciated by consumers. The results showed that OFI cladodes have high moisture and a long shelf life and are appreciated by the larvae that prefer them to other matrices. Thus, OFI can replace the conventional wet source in TM rearing, at least in areas where this cactus grows wild and is not difficult to obtain.

Keywords: *Tenebrio molitor* rearing; *Opuntia ficus indica*; alternative protein source; human food; novel foods

1. Introduction

The world's challenge today is to feed the entire human population sustainably. In fact, the increase in the population to over 9 billion by 2050 [1] and the consequent increase in the demand for food will lead to the excessive exploitation of animals and habitats, with deforestation and increased greenhouse gas (GHG) emissions [2]. In this context, the food system releases more than a quarter of all GHG emissions, most of which are due to livestock [3].

Reducing the consumption of emissions-intensive animal-based foods can improve our dramatic environmental and climatic scenario. However, the trend towards plant-based diets cannot assure either health results or higher sustainability [4]. There is a need to find alternative, more sustainable protein sources—best if produced using a circular economy perspective.

Given their biological characteristics and nutritional profile, insects lend themselves to being a solution to global food shortages [5,6]. In nature, about 2000 species of edible insects can provide food for both humans and farm animals such as fish, poultry, etc. [7].

Among edible insects, *Tenebrio molitor* (TM), a beetle of the Tenebrionidae family, also named yellow mealworm, has been extensively studied in the last few years in scientific circles as an alternative source of protein, especially since the European Commission approved it as a Novel Food and deemed it fit for human consumption in June 2021 [8].

TM has several advantages: it has a higher efficiency in converting food to body weight than meat animals, it reproduces and grows quickly, and its edible fraction is almost 100% compared to chicken and pigs (55%) and cattle (40%) [9]. It is currently widely used for animal feed such as fish, birds, and poultry, both as a whole larva and as a processed product (protein meals, protein extracts, and oils) [10,11].

TM rearing is much more sustainable than traditional livestock farming, as it emits fewer GHGs and requires less space and water [12]. Moreover, TM fits perfectly into a circular economy logic as it can grow and efficiently convert agro-industrial by-products into protein, thus valorizing secondary feedstocks and reducing food loss and waste [13].

TM can exploit organic matter as a source of nutrients and water. Water supply is crucial in insect rearing. It is known that TM larvae (TMLs), unlike adults [14], can recover water from atmospheric moisture, but only at high levels (88–92%) [15]. Such values, on the other hand, positively influence larval growth, causing their weight to increase already in the early period of development [16].

Humidifying the rearing environment, however, is not energetically sustainable as it requires additional water and, in the case of facilities like climatic chambers (especially on large-scale farms), high energy consumption. Moreover, highly humid environments favor the growth of mold and bacteria.

The most feasible alternative is to provide water using discrete doses of vegetables (potatoes, carrots, pumpkin, red cabbage, and red beet) distributed 2–3 times a week [17]. Such additional feeding does not harm rearing and even reduces the larval development time [9], increases the average weight of larvae of the same age [18], and decreases the feed conversion ratio (i.e., the amount of feed it takes to grow one kilogram of larvae) [19].

To ensure the economic and environmental sustainability of small- and medium-scale TM farming, it is preferable to use "water-supplying vegetable matrices" (hereinafter VMs) such as agri-food by-products that are inexpensive, seasonally available, and do not compete with human nutrition [20].

For our TM rearing, we focused on the cladodes of the prickly pear cactus (*Opuntia ficus-indica* (L.) Mill., hereinafter OFI). This cactus is a rustic crop, cultivated in warm climate areas for its fruit, but it is also very common on poor, dry soils, or marginal lands due to its easy adaptation. For non-productive purposes, it is used as a defensive hedge (due to the presence of spines) and windbreak. It is widely used in soil erosion protection and degraded area restoration programs, providing food resources during drought [21]. Mexico is the world's largest producer of prickly pears, while Italy (specifically Sicily) is the biggest European producer [22].

In specialized productive orchards, annual pruning produces a large amount of waste of cladodes/immature fruits which is usually wasteful and costly to farmers [23]. However, recent studies have shown that these by-products can be sources of high-value components with antioxidant activity, exploitable as anticancer, anti-inflammatory, hypoglycemic, lipid-lowering, and cholesterol-lowering agents [24,25].

OFI has several advantages as a water source for small- to medium-scale insect farming. It is widely spread and cultivated in the Mediterranean area and semi-arid climate areas of South Africa, Australia, and Central and South America. OFI cladodes are available throughout the year, even in semi-arid areas, and store well at room temperature, thus allowing to reduce energy costs for storage [26].

This article's principal objectives are to evaluate OFI as a replacement for the most common VMs, at least in areas where OFI is widespread, and demonstrate whether it is possible to reduce the wet supplementation frequency to reduce related labor costs. To achieve these goals, we compared the performance of OFI to other selected fresh VMs, in terms of dehydration and shelf life in the rearing chamber, the choice of TMLs, the palatability by TMLs after 6 days under rearing conditions, and the larval weight gain and survival.

2. Materials and Methods

One-year-old OFI cladodes were collected in the ENEA—Trisaia research centre (in the Basilicata region, in the south of Italy), where they grow wild. The cladodes are harvested when they reach a size of about 40 × 20 cm^2 and are brushed thoroughly to remove thorns and stored at room temperature. They were provided to TMLs with the peel because its presence protects them from molding and slows down dehydration, mainly due to their fatty acids that keep out water-soluble molecules [27].

The other tested VMs were purchased from local retail suppliers. All experiments were conducted in a rearing room and on TMLs reared at the Trisaia research centre under standard conditions of temperature (28.0 ± 0.1 °C), relative humidity (60 ± 5%), and photoperiod (0L:24D) [28]. The Larvae's standard diet consisted of 95% bran and 5% zootechnical yeast, purchased from local suppliers of livestock products. The larvae used in the various tests were reared in standard conditions and fed with the standard diet, adding slices of carrot twice a week, as a source of water, vitamins, and minerals.

2.1. Dehydration Test of Some VMs

One of the aims of this article is to look for an alternative to reduce the frequency of administration of VMs to TMLs. The VM chosen, however, must have particular characteristics, including that it does not dry out or go moldy when administered (i.e., under rearing conditions) and does not require removal within a few days of administration. For this purpose, the suitability of vegetables as a wet supplement was tested over time, under standard rearing conditions. The tested VMs are carrot (C), used as a control, OFI, potato (P), white apple (WA), red apple (RA), and red beet (RB). Specifically, 2 cm × 2 cm × 1 cm cubes of each VM, weighing on average 4.46 ± 0.41 g, without peel, were placed in the rearing chamber for 4 observation times, i.e., 3, 6, 9, and 12 days (respectively indicated as T3, T6, T9, and T12), according to a completely randomized design with 10 pieces (replicates) per treatment. Weight loss was used to estimate water loss over time. The presence of mold/rot was verified for each observation time.

In addition, for non-moldy VMs after three days, the initial percentage of water content was determined using a thermobalance (Gibertini Eurotherm, mod. EUJ; precision ± 1 mg).

2.2. Choice Test by TMLs

The 60-day-old TMLs (kept for 24 h without wet supplementation) were weighed and divided into 10 groups each of 15 larvae, balancing the weights of the replicates to avoid larger and more ravenous larvae all ending up in the same group.

The protocol for the choice tests was adapted from Morales-Ramos et al. (2020) [29]. In detail, multiple-choice arenas (12 cm-diameter Petri dishes) were set up with cubes of three VMs (OFI, C, and P), weighing enough to provide 3 g of water/cube of VMs, according to the percentages of water contained in each VM previously determined using the thermobalance. Thus, the water content, which was the same for all the matrices, could not influence the choice of larvae. The average cube weights were OFI = 2.68 ± 0.08 g, C = 2.44 ± 0.05 g, and P = 2.65 ± 0.04 g.

Fifteen larvae/arena were released simultaneously into the center of the arena (Figure 1). The number of larvae present on each VM was observed after 20 min (the maximum time defined by the preliminary tests). The test was performed under standard rearing conditions using larvae reared on a standard diet with the wet supplement in the first case consisting of carrots (larvae taken as our control, CtL) and in the second case, it consisted of various vegetable matrices (VvL). This was carried out to test whether the habit of a particular wet supplement influenced the larvae in their choice. The choice tests were repeated 10 times (replicates) for each group of larvae (CtL and VvL).

Figure 1. Choice test. Arena prepared with the OFI-P-C equally spaced from the center and 15 TMLs in the choice phase.

2.3. Palatability Test of the VMs by the TMLs

60-day-old TMLs (0.75 ± 0.09 g), kept for 24 h without wet supplementation, were used to investigate the palatability of the VMs, both fresh ("Fresh") and stored in a rearing room for 6 days ("Stored"). In detail, larvae fed on the standard diet were administered fresh portions (0.56 ± 0.07 g) and stored portions (0.17 ± 0.05 g) of the VMs (carrot (C), potato (P), and OFI). After 96 h, the residual weight of the VMs is measured to evaluate the amount of VMs eaten by the larvae. Ten replicates of each VM were conducted using ten larvae/replicate.

In addition, we calculated the percentage of moisture in the three VMs, both in the "Fresh" and "Stored" samples, by using the thermobalance (Gibertini Eurotherm, mod. EUJ; precision ±1 mg).

2.4. Performance of the TMLs Fed on the Standard Diet and the VMs

Next, 30-day-old TMLs (6.2 ± 0.5 mg), previously reared under standard conditions, were used to evaluate the effect of three VMs (carrot (C), potato (P), and OFI), added to the standard diet, on larval weight gain. Complete randomization was applied to the experimental design, with 10 replicates/treatments and adding about 1 g/replica of different VMs to the standard diet on a weekly basis. Each replicate consisted of groups of 20 larvae, according to Rumbos et al. [17].

After one month, the trial was stopped, and larval survival was evaluated. The final larval weight was determined with an analytical balance (Gibertini, mod. E42S-B; precision ±0.1 mg), and the weight gain/larva was calculated.

2.5. Statistical Analyses

For all tests, the data collected were statistically processed using the software GraphPad Prism version 8 (GraphPad Software, San Diego, CA, USA). The values are shown as the means ± standard deviation (SD). Homogeneity of variance and normality was verified prior to the statistical analysis. If this was confirmed, one-way ANOVA was used, followed by Tukey's post hoc test to identify the differences between the multiple theses. Alternatively, the Kruskal–Wallis test with pairwise multiple comparisons (Dunn's test) was applied. An unpaired t-test was used to compare the means of two groups of independent values with the homogeneity of variance and normality. Significance was assumed at $p < 0.05$.

3. Results

3.1. Dehydration Test of Some VMs

The dehydration test in the rearing chamber provided interesting data on both weight loss (Table 1) and mold and rot resistance (Figure 2) at four observation times: 3–6–9–12 days (named T3, T6, T9, and T12, respectively).

Table 1. Percentage of weight loss in the six VMs during four times of observation.

VMs	T3	T6	T9	T12
OFI	54.02 ± 5.32 [a]	78.13 ± 9.32 [a]	90.88 ± 1.03 [a]	91.22 ± 8.38 [a]
P	58.85 ± 7.02 [ab]	76.20 ± 6.25 [a]	86.32 ± 1.15 [b]	86.16 ± 1.28 [bc]
C	62.72 ± 3.45 [b]	78.74 ± 2.82 [a]	87.00 ± 1.01 [b]	85.13 ± 0.35 [b]
WA	45.18 ± 4.12 [c]	59.92 ± 5.21 [b]	81.71 ± 0.93 [c]	79.73 ± 1.04 [d]
RB	46.50 ± 2.13 [c]	63.07 ±6.17 [b]	77.17 ± 2.27 [d]	76.67 ± 1.04 [d]
RA	46.45 ± 3.48 [c]	66.27 ± 1.37 [b]	81.13 ± 0.44 [c]	84.91 ± 1.07 [c]

Notes: VMs = water-suppling vegetable matrices. Values with different letters within columns are significantly different according to Tukey's test ($p < 0.05$).

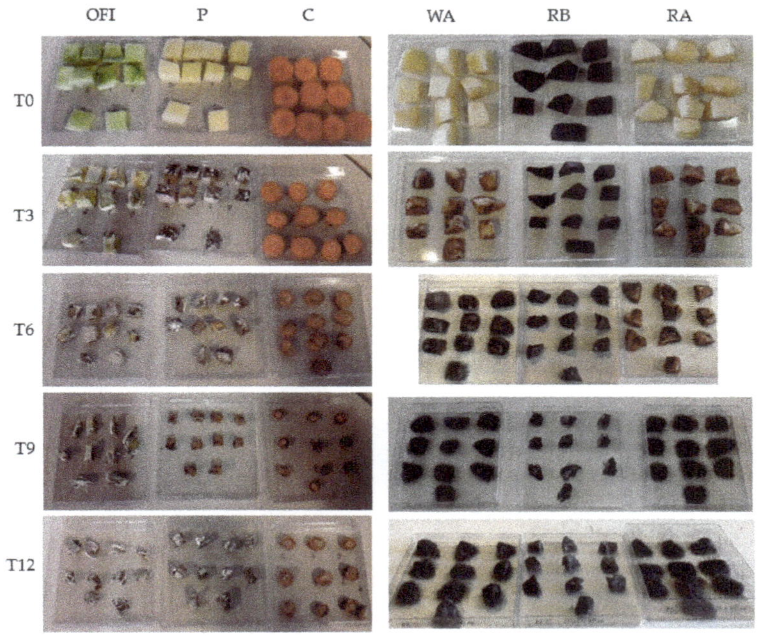

Figure 2. Evolution of the six VMs tested during the dehydration test in the rearing room under standard conditions.

Comparing the weight loss of the various matrices at the four observation times using the unpaired *t*-test ($p < 0.05$), it can be observed that the values at T9 and T12 for all the VMs are unchanged, while the values at T3, T6, and T9 are significantly different, probably because after 9 days, they have no more water to lose, and only their dry weight remains. The comparison between the results for the different VMs for each observation time using Tukey's multiple comparison tests shows that at T3, OFI is significantly different from all matrices except P, and C is different from every other VM except P. At T6, OFI, P, and C are significantly different from WA, RB, and RA. At T9, P equals C and WA equals RA. Finally, at T12, OFI is still significantly different from all the matrices.

What happens during the dehydration test in the rearing chamber to two of the six VMs is evident in Figure 2. The white apple (WA) and the red apple (RA) develop mold after only 6 days and show clear signs of rotting already at 3 days; their texture also becomes immediately too soft, making them very difficult to handle.

At T3, clear signs of desiccation also begin to appear on the potato samples. The other 3 VMs (OFI, C, and RB), although showing small changes, are still usable at T6. At 9 days (T9), OFI and C also show clear signs of desiccation, while, apparently, RB seems to resist better, probably because it has a higher dry residue and lower water content than the other three VMs at the start (Table 2). At T12, all matrices are strongly altered and their use as a source of water is certainly not recommended.

Table 2. Percentage of dry matter and moisture in the four VMs calculated by the thermobalance.

VMs	Dry Matter (%)	Moisture Content (%)
Carrot (C)	12.37 ± 0.32	87.63 ± 2.19
Potato (P)	18.06 ± 0.51	81.94 ± 2.05
Prickly Pear (OFI)	11.07 ± 0.28	88.93 ± 2.22
Red Beet (RB)	31.15 ± 0.84	68.85 ± 1.91

Note: VMs = water-suppling vegetable matrices.

Although RB has shown excellent resistance to dehydration, given its lower water content than other VMs, a larger amount of RB should be added to the standard diet to guarantee the appropriate water supplement, or it should be added twice a week, as is usually carried out with carrots. In addition, it is not a convenient alternative wet supplement, because of its cost and use for human nutrition. Therefore, we ruled it out in subsequent tests.

3.2. Choice Test by TMLs

At the end of the choice test, the number of larvae on the three VMs was detected, both for larvae that received carrots (CtL) and various vegetable matrices (VvL) as their water source.

The unpaired *t*-test shows that the choice of VM is not influenced by the type of wet supplement habitually consumed, as there is no statistically significant difference between the values in the same column, as shown in Table 3.

Table 3. Number of TMLs on the three VMs.

	OFI	P	C	No Choice
CtL	6.7 ± 1.4^{a}	2.7 ± 1.1^{b}	2.5 ± 1.1^{b}	3.2 ± 1.6^{b}
VvL	5.8 ± 1.9^{a}	2.8 ± 1.1^{b}	2.6 ± 1.1^{b}	3.8 ± 2.3^{b}

Notes: CtL = number of larvae habitually fed with a standard diet and carrot (control); VvL = number of larvae usually fed on a standard diet and various vegetable sources. Values with different letters within rows are significantly different according to Tukey's test ($p < 0.05$). The unpaired *t*-test for the comparison between the values in the same column shows no significant differences.

3.3. Palatability Test of the VMs by the TMLs

In this test, we observed both groups of larvae ("Fresh" and "Stored") for 96 h and noticed that as soon as we provided the 'Fresh' VMs, the larvae pounced on them, whereas as soon as we provided the 'Stored' VMs, the larvae took a while to approach the VMs.

The results on the VMs' consumption are shown in Figure 3.

The data were analyzed using an unpaired *t*-test to detect differences between the types of administration for each VM (Figure 3a) and Dunn's test to evaluate differences between the VMs by the type of administration (Figure 3b).

Figure 3a shows no significant difference between the consumption percentages of 'fresh' and 'stored' OFI. In contrast, the percentages of consumption between the 'Fresh' and 'Stored' matrices are significantly different for both P and C. Figure 3b shows that among the "Fresh" VMs, the consumption percentage of OFI is different from that of P, while C is not different from the other two. Furthermore, among the "Stored" VMs, the consumption percentage of C is significantly different from that of OFI and P.

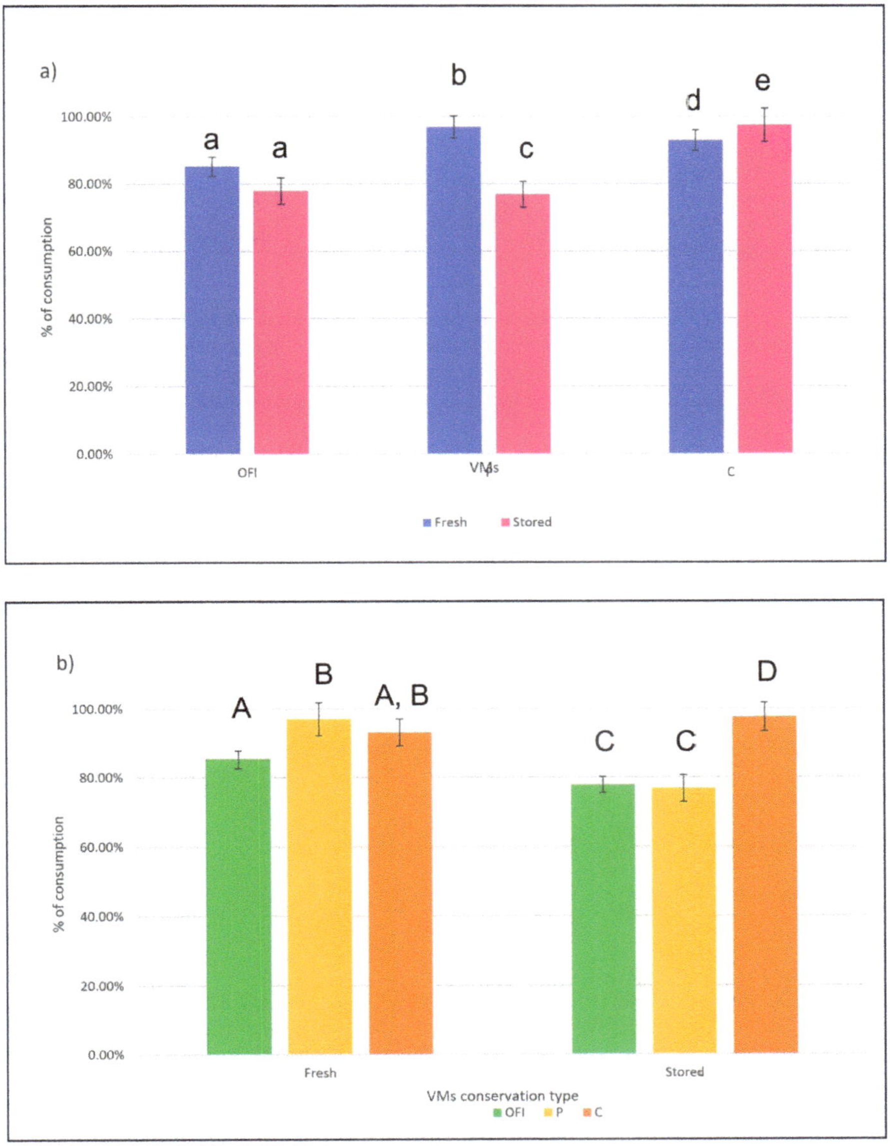

Figure 3. Percentage of VMs consumed by the TMLs after 96 h. (**a**) Percentage of consumption for each VM, administered as "fresh" (blue bars) and "stored" (pink bars). Bars with different lowercase letters are significantly different according to the unpaired *t*-test ($p < 0.05$). (**b**) Percentage consumption of the three tested VMs for each administrative condition. Bars with different uppercase letters are significantly different according to Dunn's test ($p < 0.05$).

The analysis of the three VMs with the thermobalance provided interesting results (Table 4).

Table 4. Percentage of moisture in the three VMs calculated by the thermobalance. The analyses were conducted on both fresh samples ("Fresh") (see Table 2) and those stored in a rearing room for 6 days ("Stored").

VMs	Moisture (%)	
	Fresh	Stored
Carrot (C)	87.63 ± 2.19 [ac]	29.66 ± 0.86 [a]
Potato (P)	81.94 ± 2.05 [b]	28.60 ± 0.73 [a]
Prickly Pear (OFI)	88.93 ± 2.22 [c]	50.27 ± 1.51 [b]

Note: Values with different lowercase letters within columns are significantly different according to Tukey's test ($p < 0.05$).

Tukey's test for the comparison between the moisture data for the three VMs shows that the three fresh samples have similar percentages of moisture content, except for P, which has a slightly lower percentage. After 6 days, the moisture percentage is similar only for P and C, while OFI has approximately twice as much moisture as the other two matrices.

3.4. Performance of the TMLs Fed on the Standard Diet and the VMs

The weekly supply of the tested VMs, as a moist supplement to the standard diet, ensured a larval survival rate of 98% (Table 5). Furthermore, after one month of observation, larval survival was not significantly different between the treatments. Conversely, the weight gain/larva was differently influenced by the provided VM (F = 34.907; df = 3.36; $p \leq 0.0001$). In fact, the weight gain in the P-fed larvae was significantly less than in the other treatments. In the C-fed larvae and OFI-fed larvae, the weight gain/larva was significantly higher than in the P-fed larvae, but they were not different from each other.

Table 5. Effects of different VMs on larval survival and weight gain; VMs were added as a wet supplement to the standard diet for one month of rearing.

Wet Supplement	Survival (%)	Weight Gain/Larva (mg)
Potato (P)	98.5 ± 3.4	53.8 ± 4.2 [a]
Carrot (C)	98.0 ± 3.5	62.5 ± 2.1 [b]
Prickly Pear (OFI)	98.0 ± 3.5	64.8 ± 3.5 [b]

Notes: Survival values subjected to the Kruskal–Wallis test with pairwise comparison; weight gain values subjected to one-way ANOVA followed by Tukey's post hoc test. The means with the same letter within columns are not significantly different at the 5% level of significance.

The lower weight gain in the P-fed larvae could be because we fed them the same quantity of wet supplement per week (1 g/replicate) instead of the same amount of water. Indeed, the potato contains a lower water percentage than the other two administered VMs (see Table 2).

4. Discussion

We tested OFI cladodes as a wet supplement both to test the possibility of using a more sustainable alternative in areas where it is spontaneous and widespread and to decrease the frequency of administration and, thus, the costs of managing the rearing activity.

The need to supplement the TM's dry diet with a vegetable matrix providing water and minerals, as well as some vitamins, is reported by many authors, who mostly used potato slices [29,30] once or twice a week [17] or sliced carrots [31] once a week [32]. Ruschioni et al. (2020) tried feeding olive pomace in addition to the dry diet [33].

Wet supplementation is so important in TM rearing, both in terms of survival and larval weight gain, that Morales-Ramos et al. established an optimal diet consisting of 20% of a mixed substrate, which in turn consists of 83% moist integration, i.e., more than 16% of the total diet consists of water [34]. Deruytter et al. (2021) studied the influence of wet supplement distribution on the weight gain of TMLs, by testing the different distances between wet supplement pieces and different numbers of pieces per larval density [35].

Although the comparison between TM and meat animals (pigs, chickens, and cattle) sees TM as the winner in terms of lower water consumption per gram of protein [36], the TM's water footprint can certainly be reduced by appropriately choosing the water-supplying matrix. In this sense, the same study by Miglietta et al. (2015), which focused on medium–large industrial rearing of TM that uses a diet of mixed cereals with small percentages of brewer's yeast (like our standard diet) and carrots as VMs, calculated an annual food consumption of 182 t of mixed grains and 260 t of carrots. These quantities correspond to the virtual water consumption of approximately 310,000 m^3 and 51,000 m^3, respectively. In addition, carrots need much more watering than mixed grain feed. The peculiar characteristics of OFI, a cactus with low water and cultural needs, make it ideal for significantly reducing the water footprint of medium-scale TM rearing [36].

In addition to the water footprint, the choice of the water-supplying vegetable is a critical point in increasing the whole sustainability of mealworm rearing. Coudron et al. (2022) calculated that 1.35 kg of VMs is required for each kg of worms produced. The authors pointed out that the use of VMs requires frequent procurement, energy costs for their transport and storage, and a lot of labor, which obviously affects the production costs and economic sustainability of TM rearing [37].

For this purpose, we chose to test the prickly pear cladodes because they have the necessary requisites for their valorization as by-products, are available all year round, are not subtracted from the human diet, are sufficiently inexpensive, and require little energy or economic expenditure for supply and storage [20]. From this point of view, the most frequently used matrices (potatoes, carrots, and apples) are usually by-products of the agri-food chain, which could be valorized in the same sector for juices, extracts, etc. Conversely, the cladodes are the by-product derived from pruning OFI orchards for fruit production, which is not intended for human nutrition. The absence of competition between resources and human needs is therefore underlined.

Moreover, energy consumption for storage is one of the most critical aspects for the other water-suppling vegetables. In our trials, we observed that apples rot very quickly (Figure 2). Red beets are more resistant to dehydration and molding but contain too little water (Table 2) to be a sustainable alternative, as they need to be added more frequently to the standard diet. Potatoes have a water content slightly lower than carrots and OFI and are quite palatable to TMLs. However, they show significant signs of dehydration as early as day 3 in the rearing chamber, thus requiring more frequent feeding. Finally, carrots, normally used as a control, have a water content comparable to OFI and a similar dehydration/molding pattern. However, they are less attractive to the larvae. Testing the palatability of "fresh" and 6-day-old matrices under standard rearing conditions ("stored"), we observed that TMLs like both "fresh" and "stored" OFI equally. Among the "fresh" VMs, they like the potato slightly more, whereas, among the "stored" ones, the preference is definitely for the carrot (Figure 3). What makes us favor the use of OFI is its water content after 6 days, which is about twice that of the other two VMs tested. We have also shown that providing OFI as a wet supplement to the standard diet guarantees a very good survival rate for the TMLs (equal to that provided by carrots and potatoes) and a larval weight gain comparable to that obtained by feeding them carrots (Table 5). The larval weight gain due to potato feeding is slightly lower, as expected, due to the lower water content of potatoes per gram of VMs fed. This demonstrates that the amount of water supplied to the TMLs by the wet supplement influences their weight gain and suggests that wet supplementation also provides important nutrients for larval growth.

Our results suggest that OFI represents a valid and easily available alternative VM, which is particularly attractive to TMLs. The cladodes have an interesting water content, which makes it possible to replenish the TM herds even every 10 days.

To our knowledge, this is the first time that OFI has been considered a possible wet substitute in TM rearing, or at least, there are no other scientific articles on its use in this context. For this reason, no literature data are available to compare with our results. The current study represents a novelty in introducing and evaluating OFI as a cost-effective, easily available, and more sustainable alternative VM.

On the other hand, based on our experience, using OFI may have limitations in medium-scale TM rearing, which would merit further investigation. For example, when OFI is administered in the rearing vessels, the larvae pounce on the water-rich pulp and leave the peel intact, which then dries up until the next addition. These peels end up in the leftover feed and must be removed before the new OFI administration. Another limitation could be the number of cladodes needed, which, by projecting our results for a medium-scale farm, could be around 30 kg/month. As a result, TM rearing should have an orchard or a marginal field large enough to provide such quantities.

On the other hand, based on our experience, using OFI may have limitations in medium-scale TM rearing, which would merit further investigation. For example, when OFI is administered in the rearing vessels, the larvae pounce on the water-rich pulp and leave the peel intact that dry up until the next addition. These peels add to leftover feed and must be removed before adding new OFI administration. Another limitation could be the number of cladodes needed, which, projecting our results for a medium-scale farm, could be around 30 kg/month. As a result, TM rearing should have an orchard or a marginal field large enough to provide such quantities.

All this is crucial with a view to setting up medium–small TML rearing that is increasingly sustainable, especially in anticipation of the ever-increasing use of TMLs as a source of protein in human nutrition.

5. Conclusions

The tests conducted in this study allow us to suggest the use of OFI as a water source in TM rearing, especially in the Mediterranean area, where this cactus is widespread, but its cladodes are little used for human consumption. We have demonstrated that OFI maintains its overall qualities for more days than apple and potato and, at the same time, remains palatable for mealworms (therefore, the frequency of OFI administrations is lower). OFI does not require preservation at low temperatures (therefore, there are no costs for the investment of facilities for its conservation nor costs for energy consumption), and OFI administration like supplying water increases larval growth. In addition, residual cladodes from pruning can be used, thus enhancing what would otherwise have no value and saving water that would have been necessary for producing VMs with high water demand. Based on our results, OFI is an advantageous alternative water source that could further reduce the ecological footprint of TM rearing at the small–medium production level.

Author Contributions: Conceptualization, S.E. and F.B.; methodology, S.E, P.S. and F.B.; software, P.S.; formal analysis, S.E., P.S., A.V., A.S. and F.B.; investigation, S.E., S.M., S.D. and F.B.; resources, S.E., S.M., A.V., A.S. and S.D.; writing—original draft preparation, S.E., P.S., S.M., S.D. and F.B.; writing—review and editing, S.E. and P.S.; visualization, S.E.; supervision, S.E. and P.S. All authors have read and agreed to the published version of the manuscript.

Funding: This research received no external funding.

Institutional Review Board Statement: Not applicable.

Data Availability Statement: Not applicable.

Conflicts of Interest: The authors declare no conflict of interest.

References

1. Hertog, G.P.; Wheldon, S.; Kantorova, M.; Gu, V.; Gonnella, D.; Williams, G.; Zeifman, I.; Bay, L.; Castanheira, G.; Kamiya, H.; et al. *World Population Prospects 2022: Summary of Results*; United Nations: New York, NY, USA, 2022.
2. Ordoñez-Araque, R.; Egas-Montenegro, E. Edible Insects: A Food Alternative for the Sustainable Development of the Planet. *Int. J. Gastron. Food Sci.* **2021**, *23*, 100304. [CrossRef]
3. Crippa, M.; Solazzo, E.; Guizzardi, D.; Monforti-Ferrario, F.; Tubiello, F.N.; Leip, A. Food Systems Are Responsible for a Third of Global Anthropogenic GHG Emissions. *Nat. Food* **2021**, *2*, 198–209. [CrossRef] [PubMed]

4. Macdiarmid, J.I. The Food System and Climate Change: Are Plant-Based Diets Becoming Unhealthy and Less Environmentally Sustainable? *Proc. Nutr. Soc.* **2022**, *81*, 162–167. [CrossRef] [PubMed]
5. Patel, S.; Suleria, H.A.R.; Rauf, A. Edible Insects as Innovative Foods: Nutritional and Functional Assessments. *Trends Food Sci. Technol.* **2019**, *86*, 352–359. [CrossRef]
6. Meyer-Rochow, V.B. Can Insects Help to Ease the Problem of World Food Shortage? *Search* **2014**, *6*, 261–262.
7. Imathiu, S. Benefits and Food Safety Concerns Associated with Consumption of Edible Insects. *NFS J.* **2020**, *18*. 1–11. [CrossRef]
8. European Commission. *Commission Implementing Regulation (EU) 2021/882*; European Commission: Brussels, Belgium, 2012.
9. Oonincx, D.G.A.B.; Van Broekhoven, S.; Van Huis, A.; Van Loon, J.J.A. Feed Conversion, Survival and Development, and Composition of Four Insect Species on Diets Composed of Food by-Products. *PLoS ONE* **2015**, *10*, e0144601. [CrossRef]
10. Hong, J.; Han, T.; Kim, Y.Y. Mealworm (*Tenebrio molitor* Larvae) as an Alternative Protein Source for Monogastric Animal: A Review. *Animals* **2020**, *10*, 2068. [CrossRef]
11. Errico, S.; Spagnoletta, A.; Verardi, A.; Moliterni, S.; Dimatteo, S.; Sangiorgio, P. *Tenebrio molitor* as a Source of Interesting Natural Compounds, Their Recovery Processes, Biological Effects, and Safety Aspects. *Compr. Rev. Food Sci. Food Saf.* **2022**, *21*, 148–197. [CrossRef]
12. Smetana, S.; Palanisamy, M.; Mathys, A.; Heinz, V. Sustainability of Insect Use for Feed and Food: Life Cycle Assessment Perspective. *J. Clean. Prod.* **2016**, *137*, 741–751. [CrossRef]
13. Derler, H.; Lienhard, A.; Berner, S.; Grasser, M.; Posch, A.; Rehorska, R. Use Them for What They Are Good at: Mealworms in Circular Food Systems. *Insects* **2021**, *12*, 40. [CrossRef]
14. Cortes Ortiz, J.A.; Ruiz, A.T.; Morales-Ramos, J.A.; Thomas, M.; Rojas, M.G.; Tomberlin, J.K.; Yi, L.; Han, R.; Giroud, L.; Jullien, R.L. *Insect Mass Production Technologies*; Academic Press: Cambridge, MA, USA, 2016; ISBN 9780128028568.
15. Hansen, L.L.; Ramløv, H.; Westh, P. Metabolic Activity and Water Vapour Absorption in the Mealworm *Tenebrio molitor* L. (Coleoptera, Tenebrionidae): Real-Time Measurements by Two-Channel Microcalorimetry. *J. Exp. Biol.* **2004**, *207*, 545–552. [CrossRef]
16. Johnsen, N.S.; Andersen, J.L.; Offenberg, J. The Effect of Relative Humidity on the Survival and Growth Rate of the Yellow Mealworm Larvae (*Tenebrio molitor*, Linnaeus 1758). *J. Insects Food Feed* **2021**, *7*, 311–318. [CrossRef]
17. Rumbos, C.I.; Bliamplias, D.; Gourgouta, M.; Michail, V.; Athanassiou, C.G. Rearing *Tenebrio molitor* and *Alphitobius diaperinus* Larvae on Seed Cleaning Process Byproducts. *Insects* **2021**, *12*, 293. [CrossRef]
18. Liu, C.; Masri, J.; Perez, V.; Maya, C.; Zhao, J. Growth Performance and Nutrient Composition of Mealworms (*Tenebrio molitor*) Fed on Fresh Plant Materials-Supplemented Diets. *Foods* **2020**, *9*, 151. [CrossRef]
19. Fasce, B.; Ródenas, L.; López, M.C.; Moya, V.J.; Pascual, J.J.; Cambra-López, M. Nutritive Value of Wheat Bran Diets Supplemented with Fresh Carrots and Wet Brewers' Grains in Yellow Mealworm. *J. Insect Sci.* **2022**, *22*, 7. [CrossRef]
20. Sangiorgio, P.; Verardi, A.; Dimatteo, S.; Spagnoletta, A.; Moliterni, S.; Errico, S. Valorisation of Agri-Food Waste and Mealworms Rearing Residues for Improving the Sustainability of *Tenebrio molitor* Industrial Production. *J. Insects Food Feed* **2022**, *8*, 509–524. [CrossRef]
21. Gouhis, F.; Louhaichi, M.; Nefzaoui, A. Cactus (*Opuntia ficus-indica*) Utilization for Rehabilitating Rangelands in Arid Regions of Tunisia. *Acta Hortic.* **2019**, *1247*, 95–101. [CrossRef]
22. Palmieri, N.; Suardi, A.; Stefanoni, W.; Pari, L. *Opuntia ficus-indica* as an Ingredient in New Functional Pasta: Consumer Preferences in Italy. *Foods* **2021**, *10*, 803. [CrossRef]
23. Procacci, S.; Bojórquez-Quintal, E.; Platamone, G.; Maccioni, O.; Vecchio, V.L.; Morreale, V.; Alisi, C.; Balducchi, R.; Bacchetta, L. *Opuntia ficus-indica* Pruning Waste Recycling: Recovery and Characterization of Mucilage from Cladodes. *Nat. Resour.* **2021**, *12*, 91–107. [CrossRef]
24. Albergamo, A.; Potortí, A.G.; Di Bella, G.; Amor, N.B.; Lo Vecchio, G.; Nava, V.; Rando, R.; Ben Mansour, H.; Lo Turco, V. Chemical Characterization of Different Products from the Tunisian *Opuntia ficus-indica* (L.) Mill. *Foods* **2022**, *11*, 155. [CrossRef] [PubMed]
25. Silva, M.A.; Albuquerque, T.G.; Pereira, P.; Ramalho, R.; Vicente, F.; Oliveira, M.B.P.P.; Costa, H.S. *Opuntia ficus-indica* (L.) Mill.: A Multi-Benefit Potential to Be Exploited. *Molecules* **2021**, *26*, 951. [CrossRef] [PubMed]
26. Sipango, N.; Ravhuhali, K.E.; Sebola, N.A.; Hawu, O.; Mabelebele, M.; Mokoboki, H.K.; Moyo, B. Prickly Pear (*Opuntia* spp.) as an Invasive Species and a Potential Fodder Resource for Ruminant Animals. *Sustainability* **2022**, *14*, 3719. [CrossRef]
27. Andreu-Coll, L.; Cano-Lamadrid, M.; Sendra, E.; Carbonell-Barrachina, Á.; Legua, P.; Hernández, F. Fatty Acid Profile of Fruits (Pulp and Peel) and Cladodes (Young and Old) of Prickly Pear [*Opuntia ficus-indica* (L.) Mill.] from Six Spanish Cultivars. *J. Food Compos. Anal.* **2019**, *84*, 103294. [CrossRef]
28. Ribeiro, N.; Abelho, M.; Costa, R. A Review of the Scientific Literature for Optimal Conditions for Mass Rearing *Tenebrio molitor* (Coleoptera: Tenebrionidae). *J. Entomol. Sci.* **2018**, *53*, 434–454. [CrossRef]
29. Morales-Ramos, J.A.; Rojas, M.G.; Kelstrup, H.C.; Emery, V. Self-Selection of Agricultural by-Products and Food Ingredients by *Tenebrio molitor* (Coleoptera: Tenebrionidae) and Impact on Food Utilization and Nutrient Intake. *Insects* **2020**, *11*, 827. [CrossRef]
30. Mlček, J.; Adámek, M.; Adámková, A.; Matyáš, J.; Bučková, M.; Mrázková, M.; Vícha, R.; Vychodil, R.; Knížková, I.; Volek, Z. Feed Parameters Influencing the Breeding of Mealworms (*Tenebrio molitor*). *Sustainability* **2021**, *13*, 12992. [CrossRef]
31. Rovai, D.; Ortgies, M.; Amin, S.; Kuwahara, S.; Schwartz, G.; Lesniauskas, R.; Garza, J.; Lammert, A. Utilization of Carrot Pomace to Grow Mealworm Larvae (*Tenebrio molitor*). *Sustainability* **2021**, *13*, 9341. [CrossRef]

32. Kröncke, N.; Benning, R. Self-Selection of Feeding Substrates by *Tenebrio molitor* Larvae of Different Ages to Determine Optimal Macronutrient Intake and the Influence on Larval Growth and Protein Content. *Insects* **2022**, *13*, 657. [CrossRef]
33. Ruschioni, S.; Loreto, N.; Foligni, R.; Mannozzi, C.; Raffaelli, N.; Zamporlini, F.; Pasquini, M.; Roncolini, A.; Cardinali, F.; Osimani, A.; et al. Addition of Olive Pomace to Feeding Substrate Affects Growth Performance and Nutritional Value of Mealworm (*Tenebrio molitor* L.) Larvae. *Foods* **2020**, *9*, 317. [CrossRef]
34. Morales-Ramos, J.A.; Rojas, M.G.; Shapiro-Llan, D.I.; Tedders, W.L. Use of Nutrient Self-Selection as a Diet Refining Tool in *Tenebrio molitor* (Coleoptera: Tenebrionidae). *J. Entomol. Sci.* **2013**, *48*, 206–221. [CrossRef]
35. Deruytter, D.; Coudron, C.L.; Claeys, J. The Influence of Wet Feed Distribution on the Density, Growth Rate and Growth Variability of *Tenebrio molitor*. *J. Insects Food Feed* **2021**, *7*, 141–149. [CrossRef]
36. Miglietta, P.P.; De Leo, F.; Ruberti, M.; Massari, S. Mealworms for Food: A Water Footprint Perspective. *Water* **2015**, *7*, 6190–6203. [CrossRef]
37. Coudron, C.L.; Deruytter, D.; Claeys, J. The Influence of Wet Feed PH on the Growth of *Tenebrio molitor* Larvae. *Sustainability* **2022**, *14*, 7841. [CrossRef]

agriculture

Review

Hermetia illucens (Diptera: Stratiomyidae): Need, Potentiality, and Performance Measures

Anjani Nayak [1], Martin Rühl [1,2] and Patrick Klüber [2,*]

[1] Institute of Food Chemistry and Food Biotechnology, Justus Liebig University, 35392 Giessen, Germany; martin.ruehl@ime.fraunhofer.de (M.R.)
[2] Branch for Bioresources, Fraunhofer Institute for Molecular Biology and Applied Ecology (IME), 35392 Giessen, Germany
* Correspondence: patrick.klueber@ime.fraunhofer.de; Tel.: +49-641-9721-9289

Abstract: The research on black soldier fly (BSF; *Hermetia illucens* L.; Diptera: Stratiomyidae) rearing is on the rise. The larval ability to grow on organic substances makes it an ideal candidate for the bioconversion of agricultural and other organic side streams. While there are several publications on the variables influencing the growth and development of different stages of BSF, juxtaposing the results could be amiss. This is because of the different experimental approaches and units used by the researchers. A few publications also lack information that might be necessary for comparing the results when using similar substrate and rearing conditions. In this review, we have analyzed the studies on rearing variables such as the type of feeding substrate, substrate depth and aeration, substrate temperature, substrate moisture, pH, feeding rate, and larval density mainly, but not exclusively, for the larvae. For the adults, factors such as the cage size, fly density, light, ambient temperature, and relative humidity are considered. In addition, larval performance when fed with side streams is encapsulated. This provides a backbone for future researchers to identify the already assessed variables along with their range and encourages them to define and use standardized rearing practices for a better comparison of the results.

Keywords: black soldier fly; *Hermetia illucens*; insect rearing; rearing variables; side streams; circular economy; sustainability; insects as feed; alternative protein; waste reduction

1. Introduction

In mimicking nature, insects can be deployed to convert low-value side streams into protein-rich insect-derived feed for animals [1]. Many industrial side streams are inedible by humans and cannot be applied as feed for conventional farmed animals but could be a source of feed for rearing insects [2]. The agricultural side streams have the potential to be used as feed for insects creating a sustainable circular economy [2,3]. The insects can then be directly fed or processed and fed to animals [4,5]. Insects are a valuable source of protein and fat for both humans and animals [1,6,7]. The additional advantages of using insects are [1,4] their rich nutritional profile, which meets the amino acid requirements for humans [8], fishes [4], and terrestrial animals [9], their lower GHG production [10], a lesser land requirement because of vertical farming, a shorter generation time, a lesser water consumption [6], a higher feed conversion efficiency, and their ability to thrive on side streams [11].

In the EU, insects have been allowed as protein sources for aquaculture since 2017 [12], and for poultry and pigs since 2021 [13]. The insects include the black soldier fly (BSF; *Hermetia illucens*; Diptera: Stratiomyidae), yellow mealworm (*Tenebrio molitor*; Coleoptera: Tenebrionidae), common housefly (*Musca domestica*; Diptera: Muscidae), lesser mealworm (*Alphitobius diaperinus*; Coleoptera: Tenebrionidae), house cricket (*Acheta domesticus*; Orthoptera: Gryllidae), banded cricket (*Gryllodes sigillatus*; Orthoptera: Gryllidae), field cricket (*Gryllus assimilis*; Orthoptera: Gryllidae), and silkworm (*Bombyx mori*; Lepidoptera:

Bombycidae). Insects are categorized as farmed animals and their feeding is subject to the same laws as conventional livestock. In the EU, some of these insects have also been proposed for human food. Albeit stringent regulation, the EU approved four of these insects; in the order of approval, the yellow mealworm, house cricket, migratory locust (*Locusta migrotaria;* Orthoptera: Acrididae), and more recently, the lesser mealworm.

In this review, the BSF is considered a potential insect for mass production. Through better knowledge of feed composition and the resulting larval composition, the nutrient profile of the larvae could be modified according to the requirements of feed production [14]. BSF is one of the seven insects for which larvae have been proposed to the European Commission for use in human food, although it has not yet been approved for such use. BSF is a polyphagous insect. Additionally, BSF larvae are not as susceptible to pathogens compared to other insects used for industrial rearing [15]. All the above-mentioned traits make BSF larvae the best candidate insect for industrial production.

The BSF industry plays an important role in feed safety, animal husbandry, garbage disposal, and environmental protection [1]. However, the mass production of insects, as a relatively new industry, has some challenges both in breeding and producing high-quality insects at low cost. The major problems that still exist are upscaling, unknown diet-specific feeding rate to reduce internal competition, lack of precise information on juvenile and later instar densities to maintain optimum temperature range for larval development, and preventing any disease outbreaks. For mass production, the major setbacks are automation and digitalization, the regulations that limit the use of certain feeds for insects, and their restricted sales [16]. Although protocols are available for industrial BSF breeding and rearing [11,17], there is a lack of information on handling the possible changes in the insect behavior on different feeding substrates. Despite scale (dimension) being an important factor that interacts with and influences the variables, the reliability of the BSF industry on the published laboratory scale studies can give unexpected results. The use of a variety of substrates that tremendously vary in terms of their properties, such as surface area to volume ratio, moisture, salinity, pH, or nutrients, makes it difficult to conclude on the best diet for the BSF larvae. In addition, the differences in the development time, fecundity, behavior towards specific environmental conditions, the yield, and the protein and fat production due to the insect strain are not well known for BSF [18]. The use of agricultural side streams, although sustainable, does not promise consistency in terms of product quality and nutritional value. This risk is higher if the agricultural side stream is seasonal and collected from several suppliers. Even with all the challenges, some progress has been made in understanding the biology of the BSF. Some of the information that is known on breeding and rearing BSF can be found in the life cycle section.

There are a variety of side streams available and tested as a feed material to rear insects and, in particular, BSF. However, an assessment of factors influencing larval production and a comparison of larval growth and development on side streams has not been summarized up to date. Therefore, this manuscript focuses on the abiotic variables essential for the optimum maintenance of all life stages. Additionally, a brief overview of the animal- and plant-derived substrates used as feed is given. The variables such as the type of feeding substrate, substrate depth and aeration, substrate temperature, substrate moisture, and pH are not applicable to adult flies in the majority of cases since they do not rely classically on feed but on their energy reserves to survive and reproduce successfully [17]. However, comprehensive information was gathered from the available literature. In addition, the volumetric factors, such as the cage size and fly density, are known to affect the fecundity, hence they are also listed in detail. Due to the overwhelming amount of BSF publications, the references cited are the most important representative studies and do not include all the data from every literature available so far. This work paves the way for the process of making BSF one of the mainstream feed sources by identifying insect feeds that are economical and environmentally friendly while highlighting different experimental approaches by researchers.

2. The Life Cycle of BSF

Adult flies, once emerged, live for approximately 9–11 days (d) when provided with no food, approximately 21 d when provided with water, and up to 50–73 d when provided with sugar water [19]. Besides the size, the sex of adult BSF can be determined by external dimorphisms, including antennal and abdominal structures (Figure 1i–m). The flies also need abundant sunlight or artificial lighting with a temperature range of 25–30 °C for mating and subsequent oviposition. The mating of the flies was observed at temperatures of 27 °C and above accompanied by bright sunlight conditions [20].

Figure 1. Life cycle and sexual dimorphisms of the antennae and abdominal genitalia of adult BSF. (**a**) Single deposited egg; (**b–g**) temporal sequence of larval instars L1–L6 (prepupa); (**h**) pupa; (**i**) full body view of adult female (left) and male (right) BSF; (**j**,**k**) dorsocranial view of male and female antennae; (**l**,**m**) ventrocaudal view of male and female genital structures (scale bars: white = 1 mm, black = 5 mm).

Mating occurs 2 d after fly emergence and up to 70% oviposition happens after 2 d of mating. Females lay approximately 120–820 eggs in clutches in dry spaces near decomposing matter using their ovipositor (Figure 1m) [21,22]. Up to 86.5% of oviposition was restricted to the noon hours (h) between 12:00 and 15:00, with a temperature range of 30–33 °C. The volatile compounds from decomposing materials [20,23], fresh fruit and vegetable waste [21,24], millet porridge mash [25], and the Gainesville diet [17] stimulate oviposition. Egg traps are used for the easy collection of eggs. The eggs turn white 12 h after laying (Figure 1a) and take 4–5 d to hatch at 24–27 °C and 70–90% relative humidity (RH) [20,26,27]. There are six larval instars (L1–L6) with the prepupa being the final one (Figure 1b–g). The first two instars are very fragile and the chances of mortality are higher during handling. Larval growth takes approximately 14–20 d under ideal climatic conditions with a nutritious diet [26,28]. Due to the larval mouthparts, ingestion is facilitated in a moist substrate with a smaller particle size. The fifth instar represents the reservoir of the highest nutrients and hence the desirable stage of harvest for animal feed. Furthermore, larvae reach their highest biomass in the fifth instar. At the prepupae stage, the feeding stops and they move out of the moist substrate to pupate in a dry and safe space. Prepupae are dark brown to dark grey (Figure 1g). The longevity of the flies is also influenced by the caloric content of prepupae, which is diet-dependent. Higher caloric content in the prepupae extends the life span of the resulting adults by 4 d, as observed in wild BSF populations [26]. The pupation period is usually approximately two [29] to three [23] weeks under ideal conditions (Figure 1h).

3. Variables Affecting BSF Rearing

3.1. Type of Feeding Substrates and Corresponding Challenges

Saprophagous insects, particularly the BSF, can consume a diverse group of organic wastes [30]. Hence, BSF can be reared on a range of substrates, such as agricultural wastes [3,31–33], animal and human remains [34,35], fish wastes [36,37], food or kitchen waste [38–43], municipal organic wastes [44–47], compost leachates [30] as well as human [48,49] and livestock feces [29,50–54], and restaurant wastes [33,55] transforming them into a protein-rich biomass and residual frass that can act as a potent biofertilizer. Despite all the available side streams, there are some limitations. All types of organic materials are not allowed by the EU legislation to prevent potential dangers like the spread of zoonotic diseases. For example, side streams such as manure, catering wastes, slaughterhouse wastes, and unsold products from supermarkets and food industries (containing meat and fish) are prohibited as feeding substrates for insects in the European Union [12]. The main reason is the transmission of zoonotic diseases, especially by prions. The restrictions in the EU began due to bovine spongiform encephalopathy, which is a neurodegenerative disease in cattle [56]. Substrates currently being considered for insect feeding are dependent on availability, regulatory frameworks, and their feasibility of use [57]. To date, BSF producers in Europe are using poultry feed or pig feed as the main feed for BSF rearing. The increased demand for feed in recent decades is compensated by the import of feed ingredients such as soybeans. To meet export demand, forests and agricultural lands in developing countries are being converted to grow more feed crops [58]. Therefore, using these high-cost feeds raises the insect production price while competing with conventional feedstock. Moreover, using high-quality feed for insects, although they can grow perfectly well on low nutrient-based feed, is a waste of resources. Identifying a sustainable feed that is nutritionally ideal for the larvae and permitted by the legislation of the respective country can therefore be challenging. Feed for insects is often conceptualized based on the observed habits and habitat behaviors for that particular species in nature or in fields, rather than considering the nutrient requirements of the diet. Although observations of their natural behaviors in nature, for example, in feed consumption, give an idea of their preferred feed, nutritional composition would assist in optimizing growth and development [59]. The substrate selection for industrial-scale insect production must be performed based on its local availability, its suitability for larval growth and development, and its positive impact on waste reduction by utilizing substrates that are usually discarded. Preferentially, using side streams as feed for BSF larvae can be profitable for the industries [60]. Substrate selection also relies on its properties and how it can be best prepared to feed insects. For instance, the substrates can be blended [61], heat treated [62], steamed [41], fermented [63–65], sterilized [66], or pretreated with probiotic microbes [67–69] before usage as feed for the larvae. Furthermore, substrate properties such as particle size, pH, and moisture retention are also to be considered to make the substrate ideal for larval growth and development. Another challenge is separating the larval biomass from the feeding substrate; this is discussed in more detail in the chapter on substrate moisture. This leads to the exploration of factors that are crucial in rearing insects and how larval and adult performance responds to relevant substrate modifications.

3.2. Substrate Depth

Substrate depth is the height of the feed in which larvae live until harvest. Substrate depth is considered to influence larval development as it can affect larval movements, oxygen availability, and intraspecific competition. Larvae fed on mixed organic waste were studied for three weeks on the effect of feed depth and larval aggregation temperature [61]. A homogenized feed with a 5 cm depth had 60% of the prepupae, including the highest weighing larvae (approximately 270 mg FM) and a waste reduction of approximately 63% dry mass (DM) in comparison to all other feed depths (10, 15, and 20 cm). A feed depth of >10 cm prevented access to all the provided feed. In addition, the larval survival rate also decreased to lower than 80% as the feed depth increased [61]. Lopes et al. (2023) tested four substrate depths (1.0–6.5 cm) and also concluded that a substrate depth of higher than 5 cm impairs material reduction and bioconversion efficiency [70]. Generally, larvae accumulate in masses resulting in an increased temperature (due to metabolism and movement) known as the aggregation temperature. This temperature was highest for the homogenized feed with a depth of 15 cm. According to Brits (2017) and Lopes et al. (2023), to obtain a maximum waste reduction and higher larval biomass, the feed has to be homogenized and fed to the larvae at a 5 cm depth [61,70]. However, a depth of 5 cm is not a good option for upscaling production because of the space needed and the huge surface, which would lead to higher evaporation. Therefore, it is recommended to adjust the feed depths to <10 cm [61]. Peng et al. (2022) concluded that for the examined substrate depths of 10–20 cm, the bioconversion efficiency of pig manure is higher at the lower depths of 10–15 cm [71]. A depth of 3 cm is suggested to be the best in terms of biomass conversion efficiency and larval yield [70]. In a study by Lalander et al. (2020), the substrate depth was between 3.2 and 8.0 cm for substrates with a moisture content of 76% and ~98%, respectively. The corresponding larval survival rates were 19% and 97%; which might be the result of synergistic interactions between depth, aeration, and substrate moisture [72].

Substrate depth could also apply to neonates [21] and pupae [73,74]. A pupation substrate is not mandatory [75] but having wood shavings [3,22,73], sand or soil [22], vermiculite [74,76], or wood chips [74] prevents desiccation and was reported to support adult emergence. For instance, the percentage of fly emergence was only approximately 88% without the use of pupation substrate and increased to approximately 97% with the use of sand, topsoil, wood shavings, and potting soil. The duration from the prepupae stage to adult emergence took 17 d without any pupation substrate and sand in comparison to approximately 15 d for topsoil, wood shavings, and potting soil [22].

The limited literature makes it difficult to identify an ideal substrate depth, especially since studies usually examine specific depth ranges under different rearing circumstances. In addition, interaction with other variables such as the moisture content, homogenization level, and aeration might play a role.

3.3. Substrate Aeration

The performance of larvae could be influenced by the aeration within the substrate which in turn depends on the stickiness [21], viscosity [77], particle size [28,78], and physical properties of the feed, such as its fibrous nature [79]. The particle size range described in the literature includes 0.1–1.5 mm [75], ~0.4 mm [33], ≤1–15 mm [80], ~1 mm [63], ~5 mm [81], 4.00–6.35 mm [78], 1–2 cm [23], and ≤3.8 cm [82]. According to Dortmans et al. (2017), the substrate has to be shredded to a particle size of at least less than 1–2 cm in diameter [23]. Larval weights were slightly higher at a 6.35 mm (25 mg DM) particle size than at 4.00 mm (~23 mg DM), when fed almond hulls for two weeks [78]. The authors explain that the increased access of the feed to microbes and reduced oxygen transport are assumed to be reasons for lower larval weight with a smaller particle size [78]. Yakti et al. (2023) found that the larval weight reduced from ~125 mg FM to ~75 mg FM as the straw particle size decreased from >3 mm to ≤1 mm [80]. Usage of a blended substrate could result in less accessibility of feed for the larvae due to high feed compactness and lower oxygen levels at the bottom of the rearing box forming anaerobic zones harboring corresponding microbiota [61]. Thus, the optimal substrate depth correlates with its density. The inclusion of low compactness components such as sawdust and wood chips as matrix elements during larval rearing may increase porosity and absorption of excess water in the feed [82]. Using pupation substrates of lower compact density (e.g., potting soil and wood shavings) is recommended in comparison to substrates such as sand with high natural compaction [22].

Dried sugar beet pulp and a mixture of middlings and distillers' dried grains adjusted to a moisture content of approximately 73% were considered for studying the BSF larvae performance. According to the authors, the higher aeration in dried sugar beet pulp compared to the mixture of middlings and distillers dried grain was due to different substrate physical properties. Especially the higher fiber content in dried sugar beet pulp resulted in a higher survival rate of ~78% than in the middling mixture (~56%) and dried distillers' grains (~28%) [79]. In contrast to the former study [79], the larvae were found to prefer a coffee pulp bed that is dense and homogenous [77]. Larval dry weight and overall yield were increased by three and five times, respectively, with increasing aeration from the 0.04–0.36 mL/g feed DM/min of substrate. However, higher aeration rates of $\geq$0.95 mL/g feed DM/min of substrate demonstrated no significant difference in the larval weight or the overall yield. According to Palma and colleagues, higher aeration possibly benefitted the microorganisms competing for resources. The reduced yield in the low aeration approaches is presumably due to the competition for the already scarce oxygen between larvae and microbes. Larval growth and development are influenced by the depth of the feed due to the formation of zones and varied aeration. It is therefore important to consider the factors of aeration and microbial activity during BSF production [83]. The medium-range aeration of 0.19–0.26 mL/g feed DM/min was used by Palma et al. (2019) without highlighting its effect on larval consumption or growth rate. In addition, the room ventilation of 1.22–1.39 m^3/kg feed FM/h by channel fans has been used for the substrates with a moisture content of 90–97.5% [78]. Ventilation can help in reducing the substrate depth build-up and obtaining a drier residue that is important for the easy separation of larvae during harvest. Although water removal through active ventilation is possible, substrate moisture of more than 90% would demand higher resources in terms of energy, space, and workload for a smaller change in outcome. Substrates with less than 80% moisture content are predicted to need less than 1 mL/g feed DM/min ventilation to obtain a frass of approximately 50% moisture [72]. The aeration within the substrate can also differ based on the types of lids used. Most [21,72,84] but not all studies [15] use an open lid with or without mesh to preserve the substrate moisture.

Several substrates are used to stimulate directional oviposition but mostly as a source of volatile compounds and not considering the substrate depth preferred by the females if allowed to lay eggs directly on those substrates. In the case of direct inoculation, eggs are usually placed on the surface of the substrate. Hence, the substrate depth here is not an applicable variable.

Active aeration is usually not a regular practice in rearing BSF. Although, there are several advantages such as avoiding the formation of anaerobic zones or easy separation of larvae during harvest, the energy consumption and substrate desiccation might increase disproportionate to the benefits.

3.4. Substrate Temperature

The temperature ranges found in the literature were between 23–30 °C for L1–L5 instars, ~22–33 °C for larvae and prepupae, and 10–30 °C for all larvae, prepupae, and pupae. These developmental periods are categorized by the description from the papers. The temperature of the feeding substrate plays an important role in the growth and development of BSF (Table 1). As an example, larvae took 1.7 and 5.2 d longer, respectively, to reach a prepupal stage at 30 °C and 36 °C in comparison to 27 °C. The 36 °C group did not support the transformation of prepupae to pupae [28]. The larval growth period in cow manure was almost two months at a substrate temperature of 22 °C [36]. The total weight gain by BSF larvae was higher (4.7, 4.3, and 11.1 g FM) at 27 °C in comparison to 23 and

32 °C for chicken, cow, and hog manure, respectively. There was no significant difference in weight gain for hog manure between 28 °C and 32 °C [85].

Table 1. Effect of temperature on BSF weight gain on a grain-based diet.

Temperature (°C)	Individual Larval Weight (mg FM)	References
24.9	125	[86]
27.0	150 [1]	[28]
27.6	175	[86]
30.0	138 [1]	[28]
32.2	125	[86]

[1] Individual prepupa weight (mg FM).

Substrate temperature is directly related to the ambient temperature. On the other hand, it is also clear that there is some interaction between feed, the substrate temperature, and larval aggregation temperature resulting in a varied larval performance for the same diet (Table 1). This indicates that it is essential to rear larvae at different temperatures for the same diet to determine the effect of temperature on the preferred larval trait. However, the substrate temperature does not remain constant during the feeding trials. At the beginning of the trials with early instar larvae, the substrate temperature is more crucial and is dependent on the ambient temperature, while the larvae generate a higher amount of excess heat as they gradually develop. Temperatures can rise to 42 °C at a room temperature of 23 °C with a high larval density [21,87]. The temperature during the decomposition of organic wastes by the larvae also differs. The highest temperature of 39 °C was recorded during the first days of an experiment for the mixture of vegetable and fruit waste, while the lowest temperature was recorded for vegetable waste alone after 20 d (28 °C) [45]. None of the larvae survived at temperatures of 10 and 42 °C. Furthermore, prepupae and pupae are vulnerable even to 40 °C. At 30 °C, the pupal development was the shortest with 8–10 d depending on the diet. The period from hatching to fly emergence can be as low as 28–31 d at 30 °C to approximately 182 d at 15 °C. The temperature range of 30–35 °C revealed the highest survival in most life stages [88]. Although the slower developmental times are not beneficial for insect mass production, it is important to study the threshold temperatures that slow down the rapid metabolism but are not life-threatening to the insect.

Hence, to obtain optimum larval, prepupal, and pupal development, the temperature is a key regulator. Furthermore, the consideration of other rearing factors, in particular larval density and substrate moisture, is necessary as they might affect the substrate temperature.

3.5. Substrate Moisture Content

Ambient or relative humidity affects substrate moisture, which is the quantity of water in the given substrate. The amount of water could be provided at once at the beginning of a feed trial, also known as initial moistening [73,79,89], or at different intervals during the experiment [50]. Initial substrate moisture in rearing BSF larvae was studied within a range between 20 and 90% by various experimenters. The optimum substrate moisture for BSF development was found to be approximately 40–60%. Substrate moisture of at least 40% is essential for growth and development [50]. A 56% increase in larval yield was observed as the substrate moisture content increased from 48 to 68% [83]. Cammack and Tomberlin (2017) postulate that optimum moisture not only affected the yield but also shortened developmental times, and increased pupation and emergence rates. The larval development was a week faster and required 25–50% less feed at 70% moisture content in comparison to 55%. There was up to a 3% higher fly emergence at 70% moisture compared to the 55% diet [76].

Ewusie et al. (2018) used substrates with an initial moisture ranging between 61 and 91% [45]. In another experiment, pre-consumer (vegetable trimmings, spent coffee and tea residues, no meat) and post-consumer (leftovers with meat) food wastes, with substrate moisture contents of 70, 75, and 80% were considered [90]. Here, the differences in wet larval weight were negligible between the different moisture levels. Larval weights at the moisture levels tested varied between 119 and 125 mg FM for pre-consumer and 143 and 161 mg FM for post-consumer waste. In contrast, the post-consumer diet produced heavier larvae (~153 mg FM) compared to the pre-consumer (~122 mg FM) diet due to the higher crude protein (CP) content in the former and not because of the difference in moisture levels. The larval growth was found to be 3–5 d faster in 80% substrate moisture compared to 70 and 75%. However, the survival rate was not affected by the moisture content [90]. Although early instar larvae are more robust than eggs (Figure 1a–c), lower percentages of moisture in the pre- and post-consumer treatments resulted in the drying of substrates, disabling larvae from obtaining enough nutrients, and thus reducing growth and development [90]. Nguyen et al. (2013) examined substrates with moisture contents between 75.2 and 96.5% but the effect of moisture on larval performance is not specified [39]. A broader moisture content range between 20 and 90% studied at 10% intervals by Fatchurochim et al. (1989) revealed differences in survival rates on moistened poultry manure [50]. The BSF survival was best and similar for the substrate moisture ranging between 40 and 60% in comparison to lower (20 and 30%) and higher ranges of moisture (70, 80, and 90%). As opposed to other publications, the 70% moisture reduced the BSF survival rate to 38.8% but the weight of the adult flies was the highest (4.4 mg DM) [50]. The survival rates between different life stages can vary depending on other factors besides rearing in the moist substrate [15]. A few studies also use structuring compounds in the feed such as sawdust [91], or wheat bran and brewers' spent grain [92]. Generally, pupae are able to survive and eclose without any pupation substrate. In spite, a study testing pupation substrates of varying moisture (dry to 150%) concluded that its use does affect prepupae mortality and formation of pupae. According to the authors, moistened substrates reduced the prepupal mortality by $\geq$88%, whereas dry pupation substrates enhanced the pupation rate by up to 9%, compared to the groups without pupation substrates [74].

Similarly, the use of substrate for the egg stage is limited unless the eggs are immediately inoculated onto the substrates [3,28,36]. The direct inoculation of eggs on substrates with 60–70% [17,28,91] may lead to reduced hatchability [17]. Providing water for flies is recommended by most studies. This is performed by either spraying water on the cages [17,45] or by keeping a water source like a wet cotton wick [76].

Substrate moisture is a crucial factor not only for BSF growth and development but also for larval harvest, especially in an industrial setting. In conventional bioconversion studies, organic wastes have been used as such without adjusting the water content to the requirements of the BSF [46]. Although this could save time, separating residues at approximately 80% moisture is considered difficult to impossible. Feeding larvae daily reduces the sieving efficiency as the FM feed is mostly moist [72]. Thus, reducing the substrate decomposition rate resulting in the emission of a foul smell because of incomplete substrate degradation. Larvae also tend to escape from the moist substrate. Usually, the substrate moisture decreases with ongoing rearing because of larval movement, mixing of the substrate, evaporation, and assimilation of nutrients. Increasing moisture content is crucial for some substrates, such as almond hulls, for breaking down the particles by larvae and associated microorganisms [83]. BSF larvae can ingest the nutrients dissolved in water with ease when the substrate moisture is maintained at approximately 70% [45].

Although substrates with higher moisture content could be fed to the larvae, the use of substrates with less than 80% moisture is recommended for proper larval development and is favorable for separating larvae from frass [72]. Egg hatching does not require a moist substrate, albeit few studies inoculate the eggs directly onto it. Therefore, an optimal moisture content cannot be suggested based on the current studies.

3.6. Substrate pH

There are only a few studies on substrate pH and the corresponding effects on larval growth and development, and almost none on the effects on hatching when eggs are placed immediately onto substrates. Therefore, the influence of pH on BSF growth has not been fully elucidated, although available studies have included pH values ranging from 2.0 to 10.0. Ma et al. (2018) showed that the performance of BSF larvae is pH-dependent [93]. While the pH of the diets differed considerably at the beginning of feeding, it was observed that pH values settled between 8.9 and 9.4 at the end of feeding (Table 2).

Table 2. Change in pH and corresponding individual larval weight on different diets.

Feed	Initial pH	Duration (d)	Final pH	Individual Larval Weight (mg FM)	References
Gainesville diet	4.0–9.5	9.7	8.9–9.4	140–150	[94]
Coffee pulp	7.6	13.0	8.9	147	[95]
Chicken feed	~5.0	22.3	7.2	240	[75]
Cottonseed press cake	~6.0	23.7	8.9	140	[75]
Dairy manure	8.2	21.0	7.3	-	[96]

Both acidic and alkaline pH values were well-regulated and resisted by BSF larvae [94]. The increase in pH may be due to the release of ammonia in substrates rich in nitrogen [85,97]. The pH seems to be affected by the feeding rates or the feed amount because the measured pH of the substrates (vegetable wastes) at lower feeding rates of 60 mg FM/larva/d remains more or less constant between 7 and 8. However, higher feeding rates such as 200 mg FM/larva/d possibly create an anaerobic condition pulling the pH to acidic values ranging between 4 and 5. The acidity was assumed to have lowered the growth performance of the larvae [43]. Interestingly, two nutritionally differing diets with pH values in the neutral (6.8) and slightly acidic (4.5) range showed no difference in pH levels within the larval gut. The pH values in the anterior, middle, and posterior midgut of larvae on both diets were approximately 5.5, 2.0, and 8.5, respectively [81].

The pH of fresh dairy manure was 8.2 [96]. Conversely, the initial pH of pig manure was 6.0–6.2, whereas that of chicken manure ranged between 7.4 and 8.2 [85]. Here, the harvested larval weight was 2.4-fold higher in hog manure than in chicken manure at 27 °C. According to the authors, the difference in pH probably affected some antimicrobial peptides influencing larval growth [85]. The pH of a diet consisting of chicken manure and dairy manure in the 2:3 ratio used by Ur Rehman et al. (2019) was 7.6 [67]. The pH of the fresh pig manure used in a choice test against a plant-based side stream diet ranged between 6 and 7, while the plant-by-product was between 3 and 4. It is unknown if the larval choice for pig manure was based on their aversion to the sensed acidity from the other feed [98].

There are several studies giving initial pH values but without further relationships between pH and life-history traits of BSF larvae [67,96,98]. Understanding the changing pH during larval rearing might help to optimize the dietary composition and feeding rates.

3.7. Feeding Rate

The feeding rate is defined as the amount of feed provided to the larvae or adult flies at a given time to support their growth, development, and maintenance. There are numerous studies focused on feeding, but only a few that address feeding rates for the successful production of BSF larvae. Larval feeding could either be performed initially in one batch [48,93,94] or at regular intervals during the experimental duration. The continuous feeding interval varies across the studies from daily feeding [76,94] to feeding once a week [84] or in other intervals [41,48,72]. Studies using continuous feeding either replace the feed completely at defined periods [84,99] or add a new feed on top of the remaining feed [48,53]. The feeding aspect also differs in terms of quantity. The feed could be provided at defined feeding rates [53,89] or ad libitum [15] and sometimes with no information on the total amount of feed [2,86]. The feeding rates included in the studies ranged from 12.5 mg FM/larva/d to approximately 1000 mg FM/larva/d (Table 3).

Table 3. Larval feeding rate in various studies.

Feeding Rate (mg FM/Larva)	Duration (d)	References
60	144–215	[53]
350	15	[79]
960	12	[94]
1000	19–21	[67]
1667	21–29	[93]
13–200 [1]	10–36	[100]
13–200 [1]	38–45	[32]
50–200 [1]	15	[61]
70–170 [1]	20	[89]
90–230 [1]	~25–30	[52]
100 [1]	29–43	[99]
100–1000 [1]	12	[48]
200 [1]	12–19	[101]
~286 [1]	-	[84]
~40–73 [1]	-	[39,40]

[1] Feeding rate in mg FM/larva/d.

For enhanced prepupal growth, the frequency of adding feed is more impactful than the amount of feed itself [48]. If the final product is a protein with the desired amino acid profile and fat, a feeding rate of 125 mg FM/larva/d or higher was recommended by Brits (2017). For a faster yield that results in early prepupae formation, i.e., 60% of prepupae within 21 d after hatching, the feeding rate can be raised to 200 mg FM/larva/d. Larval biomass did not significantly differ between both feeding rates. However, the efficiency of the conversion of digested feed is highest at the 125 mg FM/larva/d (42.2%) in comparison to ~32% for the 200 mg FM/larva/d feeding rate [61]. In a study by Myers et al. (2008), larval performance varied when cow manure was fed at different feeding rates of 90–230 mg FM/larva/d, with the highest feeding rate resulting in the greatest larval weight (~178 mg FM) and the highest prepupal fresh weight (~137 mg FM), faster larval development (~25 d), and the highest adult fresh weight (~55 mg FM) [52]. In contrast, the lowest feeding rate resulted in ~142 mg FM of larval and ~89 mg FM of prepupae, ~30 d for larval development, and ~37 mg FM of adult weight. A feeding rate of 100 mg FM/larva/d chicken feed resulted in prepupae of 48 mg FM with a substrate degradation of ~42%. The feeding rate of 200 mg FM/larva/d produced 63 mg FM prepupae with a substrate degradation of ~26% [100]. In another study, 200 larvae were fed with homogenized rice straw powder at rates of 12.5, 25, 50, 100, and 200 mg FM/larva/d. In the latter feeding rate, the prepupae weight was highest with 13.6 mg DM after 38 d of rearing compared to ~2 mg DM in the lowest feeding rate after 54 d. At 200 mg FM/larva/d, the substrate consumption was merely ~10% in comparison to the feeding rate of 12.5 mg FM with 30% substrate consumption harvested at 50% prepupal formation. This confirms that, independent of the feeding rates, the harvested weights are very small and not suitable [32]. According to Parra-Paz et al. (2015), feeding up to 163 mg DM/larva/d and maintaining a density of two larvae/cm^2 is considered ideal based on the predictions from modelling in terms of biomass gain (59 g DM/m^2/d) and waste reduction index (2.6%/d) without decreasing the pH to acidic levels [43]. Klammsteiner et al. (2021) fed larvae with oil waste at 70 mg FM/larva/d and food waste at 170 mg FM/larva/d to obtain the same organic matter as in 100 mg FM chicken feed per larva [89].

After the larvae have been fed for a certain period, insects must be harvested or separated from the remaining frass. This procedure is performed either by handpicking them [53] or sieving [79], usually depending on the substrate moisture at the end of the experiment. In addition, the harvesting also varies in terms of the developmental stage. For example, harvesting was conducted at the L5 stage [102] when the first prepupae were found [103], or at 10–100% prepupae formation [82,84,104], or at the pupae stage [105]. However, other studies predefine a harvest date that could range between 8 d [80] to 20 d [89] or allow the prepupae to self-harvest by actively crawling out of the substrate [24,96]. The emerged flies are not fed in the conventional sense but are commonly offered water to drink. However, providing substances like sucrose solution [73,106,107], moistened sugar cubes [19], honey, D-Glucose, *Spirulina* or *Chlorella* powder [108], milk powder, or bacteriological peptone [107] has a positive effect on fly longevity [106] and fecundity [107]. An experiment by Bertinetti et al. (2019) with no feed, drinking water, agar with sugar water, and a mixture of sugar, bacteriological peptone, and milk powder resulted in increased oviposition and egg weight with a protein availability of 2.5% and 9%, respectively, for the agar and milk diet. The adult longevity was less than 10 d without feeding. Giving water increased the longevity by 3 d in male and 2 d in female flies. The provision of agar or milk did not significantly change the longevity in males, whereas female longevity increased to 13 and 15 d, respectively [107]. The oviposition period was 10 d in the agar diet and 17 d in the milk-based diet. The total egg mass obtained in the no diet (532 mg FM) and water diet (~556 mg FM) were lower than in the agar (~931 mg FM) and milk (~1552 mg FM) treatments. The egg hatchability was ~82% and differed significantly from the other three treatments with approximately 75% [107]. Similar to Bertinetti et al. (2019), Nakamura et al. (2016) reported an adult life span of 9–11 d without feeding the flies. However, they observed a tremendously increased longevity of approximately 21 d for both males and females when provided water. The longevity increased to approximately 48 d in females and 73 d in males with a sugar diet [19]. Besides sugar and water, the life history traits of adults were examined by a feeding experiment with protein-rich (0.05–0.5%), carbohydrate-rich (5–50%), and a 0.5% saline solution [108]. Here, the feeding rate was 2 mL/pair/d of the corresponding solutions. The adult longevity was approximately 6 d with no feeding, 18–21 d with tap water, 16–19 d with distilled water, and 10–13 d with 0.5% NaCl. The females failed to lay eggs if they did not receive water. Feeding 5% honey increased the total egg yield to ~490 mg compared to ~321 mg in tap water, while 5% glucose led to a reduction of longevity (14 d) and egg yield (241 mg). When fed with the lowest concentrations of both microalgae powders, adult longevity decreased to 14–16 d, whereas the total egg yield did not differ compared to tap water [108].

It can be concluded that for higher adult longevity and egg mass, providing an energy source such as carbohydrates and protein is recommended rather than starving or offering only water at the adult stage.

3.8. Larval and Adult Density

It is important to ensure a consistent number of larvae per container depending on the amount of feed. The amount of feed one larva could consume varies between developmental stages (Figure 1b–f). The larvae have to be provided with enough feed to minimize intraspecific competition. However, too many larvae in a container can increase the substrate temperature to an uncomfortable level, which might result in larval escape and lower larval weight gain. The number of larvae used per unit area, and their stages as well as the container dimensions differ greatly between studies (Table 4). The lack of standardization makes the comparability of data considerably difficult.

Table 4. Density of BSF juveniles and adults in various studies.

BSF Density	Container Details	References
~0.03 larvae/cm^3	30 × 30 × 6.5 cm	[40]
~0.04 larvae/cm^3	76.5 × 56.5 × 30.5 cm	[24]
~0.08 larvae/cm^3	17.8 × 11.4 × 6.5 cm	[53]

Table 4. *Cont.*

BSF Density	Container Details	References
0.11 larvae/cm^3	21 × 27 × 16 cm	[2]
0.22 larvae/cm^3	60 × 40 × 30 cm	[79]
~0.11 larvae/cm^3	1.89 L	[82]
0.18 larvae/cm^3	5.5 L	[101]
0.2 larvae/cm^3	100 L	[82]
~1.7 larvae/cm^3	0.5 L	[22]
1.2 larvae/cm^2	-	[89]
1.45 larvae/cm^2	23 × 15 cm	[84]
2 larvae/cm^2	21 × 17 × 11 cm	[72]
6 larvae/cm^2	60 × 40 × 12 cm	[72]
2–6 larvae/cm^2	-	[43]
4.2 and 6.3 larvae/cm^2	~194–2060 cm^2	[109]
4–10 kg prepupae	1.0 × 1.0 × 2.5 m and 2.5 × 2.5 × 2.5 m	[110]
~111 flies/m^3	1.5 × 1.5 × 3.0 m	[111]
500–8500 flies/m^3	45 × 45 × 45 cm	[24]
740–3704 flies/m^3	30 × 30 × 30 cm	[112]
~741 flies/m^3	30 × 30 × 30 cm	[106]
~5080 flies/m^3	27 × 27 × 27 cm	[19]

Moreover, the age of inoculated larvae varies greatly. For example, Miranda et al. (2019) used neonates (<12 h after hatching), while other studies inoculated <1 d [53], 2 d [2], 3 d [17,99], 4 d [22,40], 5 d [45,72,78,94,113], 6 d [33,67], several day-old larvae [80,85,98,101,109,114] or even larvae older than 2 weeks after hatching [48]. It is only known that eggs and early instars are more fragile compared to later instars (Figure 1a–g) [17]. However, no recommendation on the ideal larval age for inoculation onto a feed is reported in the literature. In general, it can be stated that the influence of a feeding regime of interest is more difficult to trace with increasing age, as larvae may have been previously cultivated under different circumstances and diets.

Larval densities are reported either based on egg weight [15,63,91], larval numbers calculated based on the average weight of a given number of individuals [72,79,96], or by manually counting larvae [3,45,98]. Most publications only state the number of larvae or eggs that are inoculated onto the substrate [22,40] but not whether larvae were manually counted or weighed. When using eggs, the egg weights differed from 10 mg in 9.0 × 9.0 cm Petri dish containers [27] to 150–200 mg in 19.5 × 16.5 × 9.5 cm containers [15]. It is important to be aware of the differences in hatching rate depending on the egg clutches inoculated and the resulting stocking density.

Adult density and cage size are crucial variables because of their role in egg production [24,110]. The fly densities in various experiments ranged from ~111 [111] to 8500 [24] flies/m^3 in cages of different dimensions (Table 4). Based on the fly densities used for *M. domestica* [115], a density of ≥13,000 flies/m^3 is possible [24]. In a study examining densities of 500–8500 flies, the oviposition period was significantly shorter (11 d) at 500 flies/m^3 density compared to all the other densities (~15.9 d) [24]. The total egg mass obtained increased as the fly density per cage increased, i.e., 0.4 g at 500 flies/m^3 to 7.8 g at 8500 flies/m^3. The egg mass found in female-ratio-dominant cages was always higher but not significantly different for the densities of 500, 2500, and 4500 flies/m^3. In contrast, at 6500 flies/m^3 and 8500 flies/m^3, the egg mass collected from female-ratio-dominant cages was significantly higher (7.6 and 9.0 g) than the male-dominant cages (5.2 and 6.6 g), respectively. The egg hatching rate reduced although not significant from 97.5% at 500 flies/m^3 to 92.1% at 8500 flies/m^3. Authors recommend >6500 flies/m^3 in 45 × 45 × 45 cm cages [24]. The total number of eggs obtained (based on the egg weight of 100 eggs) were 8366, 20,772, and 42,633, respectively, for 740, 1852, and 3704 flies/m^3. The density of flies in this study did not affect adult survival [112]. According to Park et al. (2016), it is recommended to use higher densities of 8–10 kg prepupae in cages ranging between 1.0 × 1.0 × 2.5 m and

2.5 × 2.5 × 2.5 m to increase the number of eggs and their mass, instead of using lower densities (4 kg prepupae) [110].

It is hence noticeable that there is a considerable discrepancy between the studies, particularly in the implementation and information provided, and that no standard density of larvae or flies could be determined to achieve optimum yield.

3.9. Ambient Temperature and Relative Humidity

Ambient temperature is the temperature of the chamber or immediate environment, where the insects are reared at a relative humidity (RH) that is often controllable. The BSF thrive well in tropical climatic conditions at most of their life stages due to their habitat origin [116], which refers to rearing temperatures above 25 °C and a comparatively high RH. However, providing a tropical climate throughout the year in the northern hemisphere is expensive. That makes it important to know the exact conditions required at different life stages in BSF production. The ambient temperature for the larval stages corresponds to the substrate temperature (besides larval aggregation temperature) and hence most of it is mentioned in the substrate temperature chapter. In the case of other life stages, temperature is an essential variable as well. The range of temperatures tested or maintained for adult longevity, mating, oviposition, and egg hatchability was between 10 and 42 °C [20,73,88]. According to Holmes et al. (2016), at 12 °C (and 70% RH), eggs fail to hatch, while at 16 °C, the time taken for egg eclosion is 7 d longer than at 19 °C (approximately 8 d). Also, the eclosion rate at an ambient temperature of 16 °C was found to be only approximately 12% compared to 75% at 19 °C. The early instar larvae from the 16 °C ambient temperature died within 3 d. In the 19 °C conditions, the rate of adult emergence was approximately 32%, indicating that the minimum threshold temperature in rearing BSF is 19 °C [117]. The pre-oviposition time was as long as 16 d at 20 °C in contrast to just 5 d at 35 °C. However, the highest egg numbers were obtained at 30 °C. No oviposition was observed at temperatures less than 12 °C [21]. The egg viability was lower than 11% at 10, 37, 40, and 42 °C. The egg eclosion time was shorter as the temperature increased from 15 to 40 °C. The eclosion was 11.4 d faster at 35 °C than at 15 °C. The highest egg eclosion rates were at 30 and 35 °C with an average of 75% [88]. Another experiment with a temperature range of 20.5–39.8 °C observed up to 99.6% of the oviposition between the temperature ranges of 27.5–37.5 °C [20].

In general, temperatures of approximately 26–30 °C and 60–70% RH are maintained in the majority of publications. A RH of 59–82% was recorded during the bioconversion of feed by BSF larvae [45]. Holmes et al. (2012) studied a RH range between 25–70% for egg eclosion and adult emergence. The prepupal development took the longest (10.4 d) at 25% RH and shortest (9.5 d) at 70% RH. The pupal and adult mortality were 65, 23, and 2% and 84, 41, and 7%, respectively, for increasing RH of 25, 50 and 70%. The adult life span was shortest (~5 d) at 25% RH and longest (~8 d) at 70% RH [118]. Under high ambient temperatures, maintaining a higher RH range of 60–70% RH can minimize evaporation from eggs or moist feeding substrates. Up to 75% of eggs hatched in just 2.6 d at 35 °C ambient temperature and 70% RH. However, the larval development time at this temperature took 16 d, whereas the shortest larval developmental time of 13 d was achieved at 30 °C (70% RH) [88]. The extent to which RH affects larval growth performance is limited based on the few publications that focus on this factor.

The RH in different rearing setups and experiments includes a range of 25–90% for both flies and eggs. Jucker et al. (2017) maintained the flies at 25% RH [3]. According to Sheppard et al. (2002), mating and oviposition can be observed at a wide range of 30–90% RH [17]. There are no studies that explicitly examine the effect of varying RH on fly performance or egg hatching, except Holmes et al. (2012). Here, the egg eclosion was approximately 7, 20, 38, 73, and 38%, respectively, at a RH of 25, 40, 50, 60, and 70%. The time taken for egg eclosion was the longest (~131 d) at 25% RH in comparison to 71 d at 60% RH and 84 d at 70% RH [118]. An adequate RH is important to prevent egg desiccation; spraying water [75,119] or using wet tissues [21] are some of the commonly used techniques.

In particular, understanding the specifics of RH could facilitate the transport of live animals and improve the safety and efficiency of regular supplies of young larvae to small, decentralized insect farmers. Based on the results of the above-mentioned studies, an ambient temperature of approximately 28 °C and a RH of approximately 70% seems to be ideal for BSF production.

3.10. Light

Light, or photoperiod, is the number of hours in a day in which an organism receives illumination. Although the influence of photoperiod on BSF larvae is unknown, studies are maintaining specific photoperiods in larval rearing. The photoperiod tested ranges between zero [63,81,98], 10 h [107], 12 h [2,45,53,73,86,99], 14 h [22,76,117] and 16 h [27], or as natural sunlight through a greenhouse [82]. The egg hatching is not affected by light color temperature [120], nor by light intensity [24]. The photoperiod does not affect egg hatching as well [24]. Nevertheless, in some studies, eggs were light exposed for 12 h [118], 14 h [22,26,76,94], 16 h [27,52] to 24 h [39]. The reason for this is probably that the different developmental stages do not have to be spatially separated, thus saving money and space. The mating of BSF is most efficient under natural sunlight [111,121]. Illumination via artificial sources was shown to stimulate reproduction in adult BSF as well, offering a way for commercial year-round production [27,119,122,123]. For artificial illumination, various systems have been examined, including fluorescence lamps [106,119,122], quartz–iodine lamps [121], light-emitting diodes (LEDs, [106,119,122,123]), halogen lamps [119,122], and metal halide lamps [121]. The wavelengths studied were 300–885 nm [122], 332–535 nm [106], 350–450 nm and 350–2500 nm [121], 380–780 nm [24], and 400–700 nm [121] range. The light intensity range included 700–3700 lux [27], 3–800 $\mu mol/m^2/s$ [111], 40 $\mu mol/m^2/s$ [24], ~50–200 $\mu mol/m^2/s$ [121], 59 $\mu mol/m^2/s$ [119], and 300 $\mu mol/m^2/s$ [108]. Liu et al. (2020) tested four artificial light sources for mating success and egg clutches. They used a 50 W halogen lamp, a combined white LED and compact fluorescent lamp of 50 W each, a 400 W metal halide lamp, and a 20 W LED lamp that matched the visual spectral sensitivity of adult BSF. From each light source, wavelengths of 300–885 nm were maintained. The highest mating success (90%) and proportion of fertile egg clutches (91.4%) were achieved from the 20 W LED lamp. No fertile egg clutches were found for the metal halide light [122]. A 500 W quartz-iodine lamp (350–2500 nm) and a 450 W rare-earth lamp (350–450 nm) were used to study mating and oviposition in BSF. The number of mating pairs observed was 70, 40, and zero for sunlight, quartz-iodine, and rare-earth lamp, respectively [121]. Artificial light rich in ~440–540 nm is recommended for enhanced mating [123]. Ultraviolet, blue, and green light between the wavelengths 332 and 535 nm are perceived by the BSF adults, influencing their mating behavior [106]. However, Zhang et al. (2010) suggest a wavelength spectrum of 450–700 nm. The mating peak was observed at an irradiance of 110 $\mu mol/m^2/s$ [121]. Tomberlin and Sheppard (2002) recommend maintaining a minimum light intensity of 63 $\mu mol/m^2/s$, and, for optimum mating, an intensity of >200 $\mu mol/m^2/s$ [111]. The mating rate increased from 23 to 70% as the irradiance increased from 0.92 W/m^2 to 431 W/m^2 [123]. The oviposition peak was on the 17th and 13th d for sunlight versus the quartz-iodine lamp and both took 4 d for egg hatching [121]. The day of oviposition was (~16 d) with a LED of low intensity (~14 W/m^2), whereas it was approximately 13, 11, and 10 d, respectively, for an LED of high intensity (~24 W/m^2), a fluorescent tube of low intensity (~5 W/m^2), and a fluorescent tube of high intensity (~23 W/m^2) [106]. Here, the authors conclude that the higher light intensity in the cages decreases longevity because of higher energy loss due to increased flying activity. LED illumination is considered to support a higher egg hatchability within the same number of egg clutches in comparison to fluorescent tubes [106].

The photoperiod differs between the studies and was shown to have an effect on fly performance. The oviposition period is light dependent [19,24,119,121]. The experimental or rearing photoperiod ranged between 2–16 h [19], 6–18 h [24], 8–16 h [112], 9 h [121], 10 h [107], 12 h [20,61,88,106,109], 14 h [93], 15 h [73], and 16 h [119]. According to Hoc et al.

(2019), the illumination periods of 2–6 h and 12–18 h had oviposition peaks at 5 and 3 d, respectively. The lowest light duration in this study yielded the lowest egg mass (4.1 g). Although the egg hatching was not influenced by photoperiod, increasing light duration from 6 to 18 h reduced the oviposition period by 3 d [24].

Neither egg hatching, nor the development of juvenile stages of BSF has been shown to be affected by light exposure, thus saving energy. In contrast, different variables of lighting, including wavelength spectrum, photoperiod, intensity, and color temperature are crucial modulators for reproductive success.

4. Assessment of BSF Larvae on Various Feeding Substrates

Although BSF larvae are popularly known as voracious feeders of biodegradable substances, their performance in terms of survival rate, developmental time, biomass, and capacity to reduce waste differs depending on the nutritional composition of the substrate. The compilation of larval growth and development on various biodegradable materials helps to obtain an overall picture of what could be more suitable for the larvae and how they can be used as converters of biodegradable substances based on their specific needs. Although the current EU regulations restrict the use of certain side streams as feed, there is a huge potential in utilizing available side streams of animal or plant origin. These include livestock side streams such as manure, slaughterhouse waste, dairy side streams, as well as food waste or agricultural side streams consisting of vegetable and fruit production and processing side streams, seed press cakes, brewer's grain, fisheries side streams, and seed husks.

The comprehensive chapter on manure is intended to provide insight into its use as feed for BSF larvae. Other agricultural side streams and food wastes could be fed to other livestock animals [124] but manure remains unused except for its use as compost or fertilizer. Therefore, the use of manure as feed for BSF larvae can minimize the competition between different animals for feed availability. Hence, a compilation of manure studies on BSF performance is gathered in the next chapter. In addition, aquaculture (and fisheries) side streams like fish rendering and fish offal, meat and bone meal, feathers, and bedding materials are also categorized as side streams. These can be a potential source of feed for the insects.

4.1. Manure

BSF larvae are found to thrive on different manure-based substrates [40,67]. Among the feed trials on manure, the ones that are used widely are chicken manure, pig manure, and cow manure. The properties of manure seem to influence larval survival. Miranda et al. (2019) found that the BSF pupation rate was 60–80% higher in fresh chicken manure compared to 2 and 4 d aged chicken manure. In this experiment, the pupation was not observed when the BSF larvae were added to the 6 and 8 d old manure [103].

In general, feed trials sometimes include manure with additives, such as specific microbes or chabazite, a zeolite that reduces unpleasant odor by absorbing volatile compounds such as ammonia (Table 5). It can be summarized that on chicken manure survival of larvae becomes challenging if the substrate moisture is too high or if the manure is aged. The larval development time is predominantly longer when feeding manure in comparison to high-quality standard diets, as shown by Oonincx et al. (2015), where the development took 20 d on a chicken feed diet [53]. In addition, the developmental time to reach the prepupal stage in fresh chicken manure was shorter (16 d; [103]). This discrepancy could be explained either by the drying and remoistening process [53] or changes in the manure-associated microbial community and changes in the nutritional composition due to the ongoing degradation process. In most cases, larvae need approximately 20–25 d of developmental time on fresh chicken manure.

Table 5. Survival rate (%), developmental duration (d), and individual BSF weight (mg FM) on chicken manure-based diets.

Feed	Survival Rate (%)	Duration (d)	BSF Stage	BSF Weight (mg FM)	References
ChM (air-dried and remoistened to 80–90%)	0.0	NA	Larva	NA	[50]
ChM (dried and remoistened to ~66%)	82.0	144	Larva	57.0	[53]
ChM + chabazite + water	86.0	-	Prepupa	90.0	[125]
ChM + *B. subtilis* strain S19	99.3	-	Prepupa	87.0	[68]
ChM + *B. subtilis* strain S15	98.7	-	Prepupa	95.0	[68]
ChM (frozen and thawed)	98.0	-	Prepupa	78.0	[68]
ChM: Cow manure 3:2	95.0	21	Larva	90.5	[67]
ChM: Cow manure 3:2 + *Bacillus* sp. strain MRO_2	99.1	19	Larva	112.5	[67]
ChM (fresh)	-	26	Prepupa	225.0	[113]
ChM (fresh)	-	13	Larva	80.0	[69]
ChM (fresh) + *B. subtilis* strain BSF-CL	-	13	Larva	93.0	[69]

ChM—chicken manure.

The larval weight was 116 mg FM in chicken feed but only 57, 69, and 74 mg FM in chicken, pig, and cow manure, respectively. Here it should be noted that the provided feed amount was rather low (60 mg/larva). The larvae were harvested when the first prepupa was observed [53]. The use of fresh poultry manure resulted in the prepupae weighing almost 225 mg FM [113]. In contrast, the prepupae weight in the fresh chicken manure was only 53 mg FM [103]. The larvae of 93 mg FM were obtained on chicken manure and inoculated with the *Bacillus subtilis* strain BSF-CL at 1×10^9 CFU/mL (1 L of bacterial inoculation to 1000 kg manure). Without bacterial inoculation, the weight of the larva was only 80 mg FM [69]. Chicken manure yielded higher larval mass in 14 d when inoculated with *Kocurina marina*, *Proteus mirabilis*, and *Bacillus subtilis* (each had 22 mg DM) in comparison to the chicken manure without any inoculation (18 mg DM). In this study, the bacterial strains were inoculated at 1% (v/w) proportion onto 500 g chicken manure at a concentration of 1×10^8 CFU/mL [126]. These data show that microbes considered for a co-digestion process seem to be beneficial for BSF larvae (Table 5). Furthermore, BSF larvae are capable of reducing the bacterial load of manure-based substrates, as shown by Erickson et al. (2004) [85]. Here, the concentration of inoculated *E. coli* O157:H7 (10^7 CFU/g) in chicken manure was reduced to approximately 10^1 CFU/g within 3 d [85]. In contrast, no reduction of *Enterococcus* spp. was reported throughout the rearing cycle [85,127,128]. Larvae exposed to contaminated manure still contained viable *Salmonella enterica* Serovar Enteritidis after 6 d in their gut [85]. In addition, foodborne pathogens like *Bacillus cereus* could also be found in the BSF larval gut. This emphasizes proper decontamination before use as food or feed [129].

Sheppard et al. (1994) found a substrate reduction rate of up to 50% in their chicken manure management system. The amount of manure used is not specified by the authors [29]. In other studies, a total reduction rate of 75% of 300 g chicken manure [125], 35.8% of 1000 kg chicken manure [69], and 40.5% from 1000 kg chicken manure inoculated with the *Bacillus subtilis* strain BSF-CL were found [69]. The substrate reduction rate was better (54%) in *Bacillus subtilis*-inoculated chicken manure than in the control diet (49%) without bacterial inoculation [126]. Another study with *Bacillus* sp. strain MRO_2 inoculation had comparable results of 48% waste reduction from a chicken and dairy manure mix, whereas the control diet without bacterial inoculation was reduced by 42% [67].

These results indicate that the chicken manure can be managed very well using BSF. However, the substrate moisture, manure age, microbiome, and quality are some of the factors to be considered. Interestingly, no study so far highlights the effect of inoculating helpful microbes on the pathogens present in the manure.

The other widely tested manure is cow manure. The weight of larvae reared on cow manure and cow manure-based diet varies greatly (Table 6).

Table 6. Survival rate (%), developmental duration (d), stage of harvest, individual BSF weight (mg FM), and feeding rate on cow manure-based diets.

Feed	Survival Rate (%)	Duration (d)	BSF Stage	BSF Weight (mg FM)	Feeding Rate (mg FM/larva)	References
CoM	85.0	-	Prepupa	137	133	[52]
CoM	71.0	-	Prepupa	179	233	[52]
CoM	91.0	24	Larva	63	1000 [1]	[130]
CoM: SCR 1:4	98.8	21	Larva	123	1000 [1]	[130]
CoM: SCR 2:3	98.5	21	Larva	117	1000 [1]	[130]
CoM: SCR 3:2	98.4	21	Larva	112	1000 [1]	[130]
CoM (dried and remoistened)	87.8	214	Larva	~74	60	[53]
CoM	-	21	Prepupa	100	150	[36]
CoM: FO 9:1	-	21	Prepupa	140	150	[36]
CoM: FO 1:1	-	21	Prepupa	150	150	[36]

CoM—Cow manure; SCR—Soybean curd residue; FO—Fish offal. [1] Feeding rate in mg FM/larva/d.

The substrate reduction in cow manure was 26% [130] and 22% [96] on a dry matter basis. The percent waste reduction increased as the soybean curd residue replaced the cow manure [130]. According to the authors, BSF larvae reared on cow manure can be used to produce clean energy coupled with manure management since they produced approximately 16 g of BSF oil in 10 d from 1200 larvae [96]. In cow manure, the *E. coli* O157:H7 abundance was similar (~10^7 CFU/g) with or without larvae and at all three temperatures and feed regimes examined [85]. Contrastingly, the presence of 15 d old BSF larvae significantly reduced the *E. coli* O157:H7 concentration in cow manure from approximately 10^7 CFU/g to 10^1 CFU/g in 3 d [127].

Pig manure was also a preferred substrate in the BSF larvae feed trials. Larvae of all ages were observed to prefer pig manure over a plant-based side stream diet, irrespective of the feed they were previously fed with. The preference for pig manure increased as the larval age increased. Admittedly, this study only conducted a preference behavior but information regarding larval development was not examined [98]. The larval survival rate was 97% in dried and remoistened pig manure with a developmental time of 144 d [53]. A study conducted by El-Dakar et al. (2021) harvested larvae of approximately 202 mg FM from fresh pig manure in 36 d of development [113]. In contrast, Veldkamp et al. (2021) yielded larvae weighing just 37 mg FM at the time of harvest, when 10% prepupae were formed. The lower weights could be due to the use of 7 d old manure in the experiment, which was stored at 4 °C until use. Additionally, some general rearing issues might explain the lower weight obtained for all dietary groups including the chicken feed control (69 mg FM) [104]. The *E. coli* O157:H7 population increased slightly in the presence of larvae in the pig manure unlike in chicken manure. However, the total biomass obtained was still higher (11 g) in pig manure than in chicken manure (5 g) [85]. In the manure management experiment using BSF, the on-farm reduction in pig manure mass was 56% DM in 14 d [54]. The waste reduction index for pig manure was 3 g DM/d [104].

The larval performance on manures from the same animal species might differ based on the feed and health condition of that particular animal. The variations in survival rate, growth, and development of BSF larvae fed similar substrates could be due to the different experimental setups, including manure age and storage, BSF strains used, time to harvest, feeding rates, and general differences in rearing conditions. More research is necessary on the use of manure and its effects not only on larval performance but also on its further use as an animal feed with a standardized rearing protocol [131]. For example, varying factors such as the initial larval age, experimental duration, climatic conditions, and stage of harvest do not give a comparable larval yield even on a similar substrate. A publication by Bosch et al. (2019), which proposes a protocol for conducting a feed trial for larval production can be used as a template. They suggest using a standard rearing diet in addition to experimental diets [131]. There is no standardization protocol for trials with adults or parameter-specific studies on the BSF. However, providing detailed information on the experiment could help understand any existing differences in the results. Since manure is considered to have a high microbial load and studies postulate ambivalent data on pathogen reduction or accumulation after BSF treatment, further experiments are recommended. However, high larval survival rates indicate a species-appropriate rearing when manure is used as feed.

4.2. Food Waste and Agriculture Side Streams

Food wastes and agricultural side streams encompass kitchen, household, canteen, and restaurant leftovers, vegetable and fruit scraps, as well as side streams from plant- and animal-based industries. Generally, these side streams are highly heterogeneous making it hard to standardize a consistent composition. At times, this could include leftovers of both plant and animal origin. Many experiments considered food waste either in whole or in combination as a feeding substrate for BSF larvae. Food processing wastes comprise the low-value side streams from agroindustry. The most popular feed trials with food waste and agricultural side streams are brewer's spent grain, fruit pomace, seed husk, soybean side streams, spent coffee, pressed seed cakes, bread and cookie remains, fermented empty fruit bunches and corn straw, potato peels, cassava peels, etc.

On kitchen waste, the survival rate of BSF larvae was only 41% [39]. Although not very different, on canteen food waste the survival rate until the pupae stage was up to 80%. However, when larvae were fed purely on oil waste from the canteen, the whole population failed to reach the pupae stage. The inhibited mobility and respiration due to the viscous consistency of oil waste and lack of easily digestible nutrients is a plausible reason for the lower survival rate and biomass increase [89]. On agricultural side streams and side streams from the food industry, the larval survival rates were above 90% such as for apple pulp, brewer's spent grain, corn meal, chicory roots, and fruit puree. None of the larvae survived on tomato leaves [132]. Feeding rice straw at 12.5 mg FM/larva/d was found to result in only half of the larvae surviving. In contrast, up to 92% and 98% survived for the feeding rates of 100 mg FM/larva/d and 200 mg FM/larva/d, respectively [32]. That implies the possibility of overfeeding low-value substrates as the larvae are able to extract the required nutrients. The brewer's spent grain obtained from four different companies was formulated by either adding just water, brewer's yeast, or molasses. This resulted in a total of twelve different diet formulations. The diets significantly differed in their macronutrient contents but the survival rate was $\geq$85% for all treatments [105]. On spent coffee, sweet potato, and dough the survival rates were 98, 87, and 83%, respectively [133]. On cottonseed press cake, the survival rate was >99%. This suggests that the presence of anti-nutritional diet components like gossypol did not negatively affect the survival rate [75]. On fish-rendering products, the larval survival rate was just 1.5%. The possible heavy metal contents in the fish diet are considered to be detrimental to larval growth by the authors [39]. A survival rate of 89% was found on food waste with a fat content of 12% [132]. Similarly, the crude fat contents of approximately 9, 13, and 19% for vegetables, tofu side streams, and food waste diet had no hindrance on larval performance. For example, larvae weighed 179, 200, and 193 mg FM, respectively, when reared for 14 d on vegetables, tofu side streams, and food wastes [91]. Although larvae can survive on most diets, several factors and diet characteristics listed in this review affect the life history traits of larvae considerably. In addition, even if the larvae manage to stay alive the time taken for development is highly variable. On a rice straw diet at a 200 mg FM/larva/d feeding rate, the larvae reached the prepupal stage in 38 d, while lower feeding rates like 12.5 mg FM/larva/d extended the larval developmental time to 54 d [32]. At a 100 mg FM/larva/d feeding rate, the larval development time was ~25 d on diets consisting of vegetable wastes or tofu dreg [31]. The total developmental time to reach the pupal stage was 22 d on a food waste diet [89]. A developmental time of 31 d was taken on brewer's spent grain consisting of a sorghum–barley–water mix to reach the pupal stage. A development time of ~26 d was found in three diet formulations composed of malt–barley–water, sorghum–barley–brewer's yeast, and barley–water diet mixes [105].

Larval weight is an important factor in choosing the diet mixture. The larval weights varied between 3 and 226 mg FM when fed on various food wastes and agricultural side streams. On kitchen waste and fruit and vegetable wastes, the larval weight was 173 mg and 123 mg FM, respectively [39]. Another study by Nguyen and colleagues (2015) obtained a larval weight of 226 mg FM on kitchen waste. The reason could be the difference in fat content [40]. In the former kitchen waste, the fat content of the diet was only approximately 5% and, in the latter, 20%. In addition, the fat:protein ratio, as well as the amino acid and fatty acid profile, might play a role [134]. The high protein and high-fat diet obtained by mixing agricultural side streams gave an 86% survival rate in comparison to a low-protein and high-fat diet (72%) until observing the first prepupa [41]. The prepupal weight on a rice straw diet at a 200 mg FM/larva/d feeding rate was only 13.6 mg DM in 38 d [32]. Another study on rice straw resulted in 100% inhibition of larval growth [61]. The addition of restaurant wastes to rice straw by an 80:20 ratio resulted in a larval weight of approximately 49 mg DM; harvested at the 50% prepupae stage. Based on additional optimization tests, restaurant solid waste and rice straw mix represented 70:30 and 0.35% of Rid-X, a commercial product with functional microbes and enzymes able to break down cellulose, lipids, and proteins, leading to a larval weight of approximately 61 mg DM [33]. The larvae reared on solid residual fraction, a residue produced after oil extraction from restaurant waste, weighed 65 mg FM [55]. Industrial food waste and household food waste yielded final larval weights of approximately 176 mg and 65 mg FM within 9 d, respectively [132]. The unexpected low larval weight of 65 mg FM on household waste cannot be explained by the nutritional quality of the diet. However, the possible presence of some harmful substances (pesticides, cleaning chemicals, etc.) might have inhibited larval growth [132]. Although the survival rate in apple pulp was up to 95.5%, the larval weight was just ~38 mg FM. It was postulated that the combination of a pH of 3.7, lower protein content (3.4%), high crude fiber (25.7%), and cellulose (21.5%) in apple pulp inhibits larval growth. Similarly, feeding fruit puree led to lower larval weights (80 mg FM) even with a 99% survival rate [132]. Fiber-rich palm oil side streams used by Klüber et al. (2022), yielded larvae of 187 mg FM in a *Bjerkandera adusta* fermented diet in ~30 d in comparison to ~149 mg FM in a non-fermented reference diet in ~41 d. The chicken feed diet in the same experiment yielded larvae of 303 mg FM in ~22 d [63]. Chia et al. (2018) prepared 12 different diets from brewer's spent grain (sorghum, barley), brewer's yeast, and cane molasses, and the best results were obtained from the following diet mixes. The larval weight in the sorghum–barley–water diet mix was ~150 mg FM, while ~200 mg FM was yielded from a barley–water diet mix. Larvae of similar weights (~150–200 mg FM) were harvested from barley–brewer's yeast, malt–barley–brewer's yeast–molasses, and sorghum–barley–brewer's yeast–molasses diet mixes. Larvae were collected as soon as they developed into pupae [105]. Composted cocoa pod husks in combination with food waste resulted in a larval weight of 112 mg FM within 18 d. The prepupal weight of 150 mg FM was obtained from larvae fed on tofu dreg [31]. The larval weight on the cottonseed press cake was approximately 160 mg FM [75]. The individual weight based on the biomass harvested was 165 mg FM in the apple and spent grain diet. A slightly lower weight of 145 mg FM was found on a diet consisting of spent grain and banana. The use of pure fruits drastically reduced the larval weight to 88–105 mg FM for banana, apple, and a mixture thereof, in comparison to spent grain alone with 136 mg FM [135]. Larval weights were 22 mg, 164 mg, and 171 mg FM on oil waste, food waste, and chicken feed, respectively [89]. On fish render, larvae gained 143 mg FM [39] and 167 mg FM [40]. The lower larval weight in fish renders in comparison to kitchen waste (173 mg FM) could be caused by the bioaccumulation of heavy metals.

The use of organic wastes as feed for larvae also aims at a reduction in substrate volume. Substrate reduction after the larval harvest has been measured in some experiments. A substrate reduction of up to 85% was observed for canteen food waste. The percent substrate reduction was 66 and 2.4% on chicken feed and oil waste, respectively [89]. Brewer's grain and corn meal had a similar waste reduction percentage of approximately 45%. Industrial food waste had a substrate degradation of 59%. Apple pulp and household food waste had approximately an 18% substrate reduction [132]. The substrate reduction was 79% for the fruits and vegetable diet, followed by 70% for bakery, 64% for cheese, 61% for sugar beet waste, and approximately 52% for brewer's grain and yeast [37]. The percentage of substrate reduction was 74% for the apple–spent grain substrate mix. A substrate degradation of 59% was found in an apple–banana diet mix [134].

It is evident that BSF larvae are capable of thriving on various organic substrates. However, the results of these feeding studies emphasize the complexity of the larval feeding regime, particularly the availability and ratio of required nutrients, the absence of contaminants, and the provision of appropriate amounts of feed. Moreover, differences in larvae performance could be attributed to dissimilarities in rearing conditions and substrate properties in terms of particle size, and substrate moisture among other factors. The distinguishable experimental approaches by the researchers have to be standardized or at least kept in mind to make an unbiased comparison of results.

5. Conclusions

For an optimized and sustainable production of BSF as a source of food and feed, ample research on available side streams and appropriate rearing conditions is crucial. Although the earlier studies conducted on BSF rearing provide good information on their performances, it is clear that the protocols and approaches in measuring the listed rearing variables differed tremendously. These differences make it challenging to compare the results between studies but even harder to translate knowledge to an industrial application. Here, a comprehensive overview of biotic and abiotic variables to be kept in mind is highlighted. These include substrate chosen as feed and its properties, environmental conditions, and container or cage type to facilitate optimum circumstances for BSF growth and development.

According to the studies analyzed, the mortality rate of larvae grown on agricultural side streams is relatively low while their performances differ between the diets, which is promising. Combining different side streams in a way that their nutritional profiles complement each other could compensate individual deficiencies. In consequence, low-value agricultural side streams can be adjusted to the nutritional requirements of BSF larvae, creating a sustainable feeding system that contributes to a circular economy. In addition, this review also guides define the range of variables to be tested in the future; ensuring an improvement first in research and long-term for BSF production.

Author Contributions: Conceptualization, A.N., P.K. and M.R.; methodology, A.N. and P.K.; software, A.N. and P.K.; validation, A.N., P.K. and M.R.; A.N., P.K. and M.R.; investigation, A.N. and P.K.; resources, M.R.; data curation, A.N. and P.K.; writing—original draft preparation, A.N.; writing—review and editing, A.N., P.K. and M.R.; visualization, A.N. and P.K.; supervision, P.K. and M.R.; project administration, M.R. and P.K.; funding acquisition, M.R. All authors have read and agreed to the published version of the manuscript.

Funding: This research was funded by the Hermetia Baruth GmbH and the German Federal Ministry for Education and Research (BMBF) grant FKZ 031B1291B (InA) within the BioBall innovation space.

Institutional Review Board Statement: Not applicable.

Data Availability Statement: All relevant data are contained within the article.

Acknowledgments: We highly appreciate the work of the anonymous reviewers.

Conflicts of Interest: The authors declare no conflict of interest.

References

1. Van Huis, A. Potential of insects as food and feed in assuring food security. *Annu. Rev. Entomol.* **2013**, *58*, 563–583. [CrossRef] [PubMed]

2. Bava, L.; Jucker, C.; Gislon, G.; Lupi, D.; Savoldelli, S.; Zucali, M.; Colombini, S. Rearing of *Hermetia illucens* on different organic by-products: Influence on growth, waste reduction, and environmental impact. *Animals* **2019**, *9*, 289. [CrossRef] [PubMed]
3. Jucker, C.; Erba, D.; Leonardi, M.G.; Lupi, D.; Savoldelli, S. Assessment of vegetable and fruit substrates as potential rearing media for *Hermetia illucens* (Diptera: Stratiomyidae) larvae. *Environ. Entomol.* **2017**, *46*, 1415–1423. [CrossRef] [PubMed]
4. Bondari, K.; Sheppard, D.C. Soldier fly larvae as feed in commercial fish production. *Aquaculture* **1981**, *24*, 103–109. [CrossRef]
5. Van Huis, A.; Van Itterbeeck, J.; Klunder, H.; Mertens, E.; Halloran, A.; Muir, G.; Vantomme, P. *Edible Insects: Future Prospects for Food and Feed Security*; Food and Agriculture Organization of the United States (FAO): Rome, Italy, 2013; Volume 171. Available online: https://www.fao.org/3/i3253e/i3253e.pdf (accessed on 20 January 2023).
6. Food and Agriculture Organization of the United States (FAO). Food Wastage Footprint: IMPacts on Natural Resources Summary Report. Available online: https://www.fao.org/3/i3347e/i3347e.pdf (accessed on 16 March 2023).
7. Barragan-Fonseca, K.B.; Dicke, M.; Van Loon, J.J.A. Nutritional value of the black soldier fly (*Hermetia illucens* L.) and its suitability as animal feed—A review. *J. Insects Food Feed* **2017**, *3*, 105–120. [CrossRef]
8. Rumpold, B.A.; Schlüter, O.K. Potential and challenges of insects as an innovative source for food and feed production. *Innov. Food Sci. Emerg. Technol.* **2013**, *17*, 1–11. [CrossRef]
9. Veldkamp, T.; Bosch, G. Insects: A protein-rich feed ingredient in pig and poultry diets. *Anim. Front.* **2015**, *5*, 45–50.
10. Oonincx, D.G.A.B.; Van Itterbeeck, J.; Heetkamp, M.J.W.; Van Den Brand, H.; Van Loon, J.J.A.; Van Huis, A. An exploration on greenhouse gas and ammonia production by insect species suitable for animal or human consumption. *PLoS ONE* **2010**, *5*, e14445. [CrossRef]
11. Surendra, K.C.; Olivier, R.; Tomberlin, J.K.; Jha, R.; Khanal, S.K. Bioconversion of organic wastes into biodiesel and animal feed via insect farming. *Renew. Energy* **2016**, *98*, 197–202. [CrossRef]
12. Regulation (EU) 893/2017 of 24 May 2017 amending Annexes I and IV to Regulation (EC) No 999/2001 of the European Parliament and of the Council and Annexes X, XIV and XV to Commission Regulation (EU) No 142/2011 as regards the provisions on processed animal protein. *Off. J. Eur. Union* **2017**, *138*, 92–116.
13. Regulation (EU) 1372/2021 of 17 August 2021 amending Annex IV to Regulation (EC) No 999/2001 of the European Parliament and of the Council as regards the prohibition to feed non-ruminant farmed animals, other than fur animals, with protein derived from animals. *Off. J. Eur. Union* **2021**, *295*, 1–17.
14. Liland, N.S.; Biancarosa, I.; Araujo, P.; Biemans, D.; Bruckner, C.G.; Waagbø, R.; Torstensen, B.E.; Lock, E.-J. Modulation of nutrient composition of black soldier fly (*Hermetia illucens*) larvae by feeding seaweed-enriched media. *PLoS ONE* **2017**, *12*, e0183188. [CrossRef] [PubMed]
15. Klüber, P.; Müller, S.; Schmidt, J.; Zorn, H.; Rühl, M. Isolation of bacterial and fungal microbiota associated with *Hermetia illucens* larvae reveals novel insights into entomopathogenicity. *Microorganisms* **2022**, *10*, 319. [CrossRef] [PubMed]
16. Jensen, K.; Kristensen, T.N.; Heckmann, L.-H.L.; Sørensen, J.G. The potential of insects as food and feed. In *Insects as Food and Feed: From Production to Consumption*; Wageningen Academic Publishers: Wageningen, The Netherlands, 2017; Chapter 9; pp. 174–192.
17. Sheppard, D.C.; Tomberlin, J.K.; Joyce, J.A.; Kiser, B.C.; Sumner, S.M. Rearing methods for the black soldier fly (Diptera: Stratiomyidae). *J. Med. Entomol.* **2002**, *39*, 695–698. [CrossRef] [PubMed]
18. Tomberlin, J.K.; Cammack, J.A. The potential of insects as food and feed. In *Insects as Food and Feed: From Production to Consumption*; Wageningen Academic Publishers: Wageningen, The Netherlands, 2017; Chapter 12; pp. 230–243.
19. Nakamura, S.; Ichiki, R.T.; Shimoda, M.; Morioka, S. Small-scale rearing of the black soldier fly, *Hermetia illucens* (Diptera: Stratiomyidae), in the laboratory: Low-cost and year-round rearing. *Appl. Entomol. Zool.* **2016**, *51*, 161–166. [CrossRef]
20. Booth, D.C.; Sheppard, C. Oviposition of the black soldier fly, *Hermetia illucens* (Diptera: Stratiomyidae): Eggs, masses, timing, and site characteristics. *Environ. Entomol.* **1984**, *13*, 421–423. [CrossRef]
21. Warburton, K.; Hallmann, V. Processing of organic materials by the soldier fly, *Hermetia illucens*. In *Integrated Biosystems for Sustainable Development*; Warburton, K., Pillai-McGarry, U., Ramage, D., Eds.; Rural Industries Research and Development Corporation; University of Queensland: St Lucia, QLD, Australia, 2002; pp. 115–126.
22. Holmes, L.A.; Vanlaerhoven, S.L.; Tomberlin, J.K. Substrate effects on pupation and adult emergence of *Hermetia illucens* (Diptera: Stratiomyidae). *Environ. Entomol.* **2013**, *42*, 370–374. [CrossRef]
23. Dortmans, B.M.A.; Diener, S.; Verstappen, B.M.; Zurbrügg, C. *Black Soldier Fly Biowaste Processing—A Step-by-Step Guide*; Eawag: Swiss Federal Institute of Aquatic Science and Technology: Dübendorf, Switzerland, 2017. Available online: https://www.eawag.ch/fileadmin/Domain1/Abteilungen/sandec/publikationen/SWM/BSF/BSF_Biowaste_Processing_LR.pdf (accessed on 18 July 2023).
24. Hoc, B.; Noël, G.; Carpentier, J.; Francis, F.; Caparros Megido, R. Optimization of black soldier fly (*Hermetia illucens*) artificial reproduction. *PLoS ONE* **2019**, *14*, e0216160. [CrossRef]
25. Boafo, H.A.; Gbemavo, D.S.J.C.; Timpong-Jones, E.C.; Eziah, V.; Billah, M.; Chia, S.Y.; Aidoo, O.F.; Clottey, V.A.; Kenis, M. Substrates most preferred for black soldier fly *Hermetia illucens* (L.) oviposition are not the most suitable for their larval development. *J. Insects Food Feed* **2023**, *9*, 183–192. [CrossRef]
26. Tomberlin, J.K.; Sheppard, D.C.; Joyce, J.A. Selected life-history traits of black soldier flies (Diptera: Stratiomyidae) reared on three artificial diets. *Ann. Entomol. Soc. Am.* **2002**, *95*, 379–386. [CrossRef]
27. Macavei, L.I.; Benassi, G.; Stoian, V.; Maistrello, L. Optimization of *Hermetia illucens* (L.) egg laying under different nutrition and light conditions. *PLoS ONE* **2020**, *15*, e0232144. [CrossRef] [PubMed]

28. Tomberlin, J.K.; Adler, P.H.; Myers, H.M. Development of the black soldier fly (Diptera: Stratiomyidae) in relation to temperature. *Environ. Entomol.* **2009**, *38*, 930–934. [CrossRef] [PubMed]
29. Sheppard, D.C.; Newton, G.L.; Thompson, S.A.; Savage, S. A value added manure management system using the black soldier fly. *Bioresour. Technol.* **1994**, *50*, 275–279. [CrossRef]
30. Popa, R.; Green, T.R. Using black soldier fly larvae for processing organic leachates. *J. Econ. Entomol.* **2012**, *105*, 374–378. [CrossRef] [PubMed]
31. Kinasih, I.; Putra, R.E.; Permana, A.D.; Gusmara, F.F.; Nurhadi, M.Y.; Anitasari, R.A. Growth performance of black soldier fly larvae (*Hermetia illucens*) fed on some plant based organic wastes. *Hayati J. Biosci.* **2018**, *25*, 79. [CrossRef]
32. Manurung, R.; Supriatna, A.; Esynthi, R.R.; Putra, R.E. Bioconversion of rice straw waste by black soldier fly larvae (*Hermetia illucens* L.): Optimal feed rate for biomass production. *J. Entomol. Zool. Stud.* **2016**, *4*, 1036–1041.
33. Zheng, L.; Hou, Y.; Li, W.; Yang, S.; Li, Q.; Yu, Z. Biodiesel production from rice straw and restaurant waste employing black soldier fly assisted by microbes. *Energy* **2012**, *47*, 225–229. [CrossRef]
34. Tomberlin, J.K.; Sheppard, D.C.; Joyce, J.A. Black soldier fly (Diptera: Stratiomyidae) colonization of pig carrion in South Georgia. *J. Forensic Sci.* **2005**, *50*, 152–153. [CrossRef]
35. Pujol-Luz, J.R.; Da Francez, P.A.C.; Ururahy-Rodrigues, A.; Constantino, R. The black soldier fly, *Hermetia illucens* (Diptera, Stratiomyidae), used to estimate the postmortem interval in a case in Amapá State, Brazil. *J. Forensic Sci.* **2008**, *53*, 476–478. [CrossRef]
36. St-Hilaire, S.; Cranfill, K.; McGuire, M.A.; Mosley, E.E.; Tomberlin, J.K.; Newton, L.; Sealey, W.; Sheppard, C.; Irving, S. Fish offal recycling by the black soldier fly produces a foodstuff high in Omega-3 fatty acids. *J. World Aquac. Soc.* **2007**, *38*, 309–313. [CrossRef]
37. Magee, K.; Halstead, J.; Small, R.; Young, I. Valorisation of organic waste by-products using black soldier fly (*Hermetia illucens*) as a bio-convertor. *Sustainability* **2021**, *13*, 8345. [CrossRef]
38. Leong, S.Y.; Kutty, S.R.M. Characteristic of *Hermetia illucens* fatty acid and that of the fatty acid methyl ester synthesize based on upcycling of perishable waste. *Waste Biomass Valorization* **2020**, *11*, 5607–5614. [CrossRef]
39. Nguyen, T.T.X.; Tomberlin, J.K.; Vanlaerhoven, S. Influence of resources on *Hermetia illucens* (Diptera: Stratiomyidae) larval development. *J. Med. Entomol.* **2013**, *50*, 898–906. [CrossRef] [PubMed]
40. Nguyen, T.T.X.; Tomberlin, J.K.; Vanlaerhoven, S. Ability of black soldier fly (Diptera: Stratiomyidae) larvae to recycle food waste. *Environ. Entomol.* **2015**, *44*, 406–410. [CrossRef] [PubMed]
41. Oonincx, D.G.A.B.; Van Broekhoven, S.; Van Huis, A.; Van Loon, J.J.A. Feed conversion, survival and development, and composition of four insect species on diets composed of food by-Products. *PLoS ONE* **2015**, *10*, e0144601. [CrossRef] [PubMed]
42. Giannetto, A.; Oliva, S.; Riolo, K.; Savastano, D.; Parrino, V.; Cappello, T.; Maisano, M.; Fasulo, S.; Mauceri, A. Waste valorization via *Hermetia illucens* to produce protein-rich biomass for feed: Insight into the critical nutrient taurine. *Animals* **2020**, *10*, 1710. [CrossRef] [PubMed]
43. Parra-Paz, A.S.; Carrejo, N.S.; Gómez Rodríguez, C.H. Effects of larval density and feeding rates on the bioconversion of vegetable waste using black soldier fly larvae *Hermetia illucens* (L.), (Diptera: Stratiomyidae). *Waste Biomass Valorization* **2015**, *6*, 1059–1065. [CrossRef]
44. Kalová, M.; Borkovcová, M. Voracious larvae *Hermetia illucens* and treatment of selected types of biodegradable waste. *Acta Univ. Agric. Silvic. Mendel. Brun.* **2013**, *61*, 77–83. [CrossRef]
45. Ewusie, E.A.; Kwapong, P.K.; Ofosu-Budu, G.; Sandrock, C.; Akumah, A.; Nartey, E.; Teye-Gaga, C.; Agyarkwah, S.K.; Adamtey, N. Development of black soldier fly, *Hermetia illucens* (Diptera: Stratiomyidae) in selected organic market waste fractions in Accra, Ghana. *Asian J. Biotechnol. Bioresour. Technol.* **2018**, *4*, 1–16. [CrossRef]
46. Diener, S.; Studt Solano, N.M.; Roa Gutiérrez, F.; Zurbrügg, C.; Tockner, K. Biological treatment of municipal organic waste using black soldier fly larvae. *Waste Biomass Valorization.* **2011**, *2*, 357–363. [CrossRef]
47. Siddiqui, S.A.; Ristow, B.; Rahayu, T.; Putra, N.S.; Yuwono, N.W.; Nisa, K.; Mategeko, B.; Smetana, S.; Saki, M.; Nawaz, A.; et al. Black soldier fly larvae (BSFL) and their affinity for organic waste processing. *Waste Manag.* **2022**, *140*, 1–13. [CrossRef] [PubMed]
48. Banks, I.J.; Gibson, W.T.; Cameron, M.M. Growth rates of black soldier fly larvae fed on fresh human faeces and their implication for improving sanitation. *Trop. Med. Int. Health* **2014**, *19*, 14–22. [CrossRef] [PubMed]
49. Lalander, C.; Diener, S.; Magri, M.E.; Zurbrügg, C.; Lindström, A.; Vinnerås, B. Faecal sludge management with the larvae of the black soldier fly (*Hermetia illucens*) from a hygiene aspect. *Sci. Total Environ.* **2013**, *458–460*, 312–318. [CrossRef] [PubMed]
50. Fatchurochim, S.; Geden, C.J.; Axtell, R.C. Filth fly (Diptera) oviposition and larval development in poultry manure of various moisture levels. *J. Entomol. Sci.* **1989**, *24*, 224–231. [CrossRef]
51. Zhou, F.; Tomberlin, J.K.; Zheng, L.; Yu, Z.; Zhang, J. Developmental and waste reduction plasticity of three black soldier fly strains (Diptera: Stratiomyidae) raised on different livestock manures. *J. Med. Entomol.* **2013**, *50*, 1224–1230. [CrossRef] [PubMed]
52. Myers, H.M.; Tomberlin, J.K.; Lambert, B.D.; Kattes, D. Development of black soldier fly (Diptera: Stratiomyidae) larvae fed dairy manure. *Environ. Entomol.* **2008**, *37*, 11–15. [CrossRef] [PubMed]
53. Oonincx, D.; Van Huis, A.; Van Loon, J. Nutrient utilisation by black soldier flies fed with chicken, pig, or cow manure. *J. Insects Food Feed* **2015**, *1*, 131–139. [CrossRef]

54. Newton, G.L.; Sheppard, D.C.; Watson, D.W.; Burtle, G.; Dove, R. Using the black soldier fly, *Hermetia illucens*, as a value-added tool for the management of swine manure. In *Report for Mike Williams*; Director of the Animal and Poultry Waste Management Center, North Carolina State University: Raleigh, NC, USA, 2005.
55. Zheng, L.; Li, Q.; Zhang, J.; Yu, Z. Double the biodiesel yield: Rearing black soldier fly larvae, *Hermetia illucens*, on solid residual fraction of restaurant waste after grease extraction for biodiesel production. *Renew. Energy* **2012**, *41*, 75–79. [CrossRef]
56. Belluco, S.; Halloran, A.; Ricci, A. New protein sources and food legislation: The case of edible insects and EU law. *Food Secur.* **2017**, *9*, 803. [CrossRef]
57. European Food Safety Authority (EFSA). Risk profile related to production and consumption of insects as food and feed. *EFSA J.* **2015**, *13*, 4257. [CrossRef]
58. Makkar, H.P.S. Review: Feed demand landscape and implications of food-not feed strategy for food security and climate change. *Animal* **2018**, *12*, 1744–1754. [CrossRef] [PubMed]
59. Barker, D.; Fitzpatrick, M.P.; Dierenfeld, E.S. Nutrient composition of selected whole invertebrates. *Zoo Biol.* **1998**, *17*, 123–134. [CrossRef]
60. Van Huis, A. Edible insects: Challenges and prospects. *Entomol. Res.* **2022**, *52*, 161–177. [CrossRef]
61. Brits, D. Improving Feeding Efficiencies of Black Soldier Fly Larvae, *Hermetia illucens* (L., 1758) (Diptera: Stratiomyidae: Hermetiinae) through Manipulation of Feeding Conditions for Industrial Mass Rearing. Ph.D. Thesis, Stellenbosch University, Stellenbosch, Western Cape, South Africa, 2017.
62. Van Looveren, N.; Verbaet, L.; Frooninckx, L.; Van Miert, S.; Van Campenhout, L.; Van der Borght, M.; Vandeweyer, D. Effect of heat treatment on microbiological safety of supermarket food waste as substrate for black soldier fly larvae (*Hermetia illucens*). *Waste Manag.* **2023**, *164*, 209–218. [CrossRef]
63. Klüber, P.; Tegtmeier, D.; Hurka, S.; Pfeiffer, J.; Vilcinskas, A.; Rühl, M.; Zorn, H. Diet fermentation leads to microbial adaptation in black soldier fly (*Hermetia illucens*; Linnaeus, 1758) larvae reared on palm oil side streams. *Sustainability* **2022**, *14*, 5626. [CrossRef]
64. Permana, A.D.; Rohmatillah, D.D.F.; Putra, R.E.; Julita, U.; Susanto, A. Bioconversion of fermented barley waste by black soldier fly *Hermetia illucens* L. (Diptera; Stratiomyidae). *Jurnal Biodjati* **2021**, *6*, 235–245. [CrossRef]
65. Gao, Z.; Wang, W.; Lu, X.; Zhu, F.; Liu, W.; Wang, X.; Lei, C. Bioconversion performance and life table of black soldier fly (*Hermetia illucens*) on fermented maize straw. *J. Clean. Prod.* **2019**, *230*, 974–980. [CrossRef]
66. Yu, Y.; Zhang, J.; Zhu, F.; Fan, M.; Zheng, J.; Cai, M.; Zheng, L.; Huang, F.; Yu, Z.; Zhang, J. Enhanced protein degradation by black soldier fly larvae (*Hermetia illucens* L.) and its gut microbes. *Front. Microbiol.* **2023**, *13*, 1095025. [CrossRef]
67. Ur Rehman, K.; Rehman, R.U.; Somroo, A.A.; Cai, M.; Zheng, L.; Xiao, X.; Rehman, A.U.; Rehman, A.; Tomberlin, J.K.; Yu, Z.; et al. Enhanced bioconversion of dairy and chicken manure by the interaction of exogenous bacteria and black soldier fly larvae. *J. Environ. Manag.* **2019**, *237*, 75–83. [CrossRef]
68. Yu, G.; Cheng, P.; Chen, Y.; Li, Y.; Yang, Z.; Chen, Y.; Tomberlin, J.K. Inoculating poultry manure with companion bacteria influences growth and development of black soldier fly (Diptera: Stratiomyidae) larvae. *Environ. Entomol.* **2011**, *40*, 30–35. [CrossRef]
69. Xiao, X.; Mazza, L.; Yu, Y.; Cai, M.; Zheng, L.; Tomberlin, J.K.; Yu, J.; Van Huis, A.; Yu, Z.; Fasulo, S.; et al. Efficient co-conversion process of chicken manure into protein feed and organic fertilizer by *Hermetia illucens* L. (Diptera: Stratiomyidae) larvae and functional bacteria. *J. Environ. Manag.* **2018**, *217*, 668–676. [CrossRef] [PubMed]
70. Lopes, I.G.; Wiklicky, V.; Ermolaev, E.; Lalander, C. Dynamics of black soldier fly larvae composting—Impact of substrate properties and rearing conditions on process efficiency. *Waste Manag.* **2023**, *172*, 25–32. [CrossRef] [PubMed]
71. Peng, C.; Zhou, T.; Song, S.; Sun, S. Analysis and experiment of feeding material process of *Hermetia illucens* L. frass bucket wheel based on DEM. *Comput. Electron. Agric.* **2022**, *196*, 106855. [CrossRef]
72. Lalander, C.; Ermolaev, E.; Wiklicky, V.; Vinnerås, B. Process efficiency and ventilation requirement in black soldier fly larvae composting of substrates with high water content. *Sci. Total Environ.* **2020**, *729*, 138968. [CrossRef] [PubMed]
73. Lupi, D.; Savoldelli, S.; Leonardi, M.G.; Jucker, C. Feeding in the adult of *Hermetia illucens* (Diptera Stratiomyidae): Reality or fiction? *J. Entomol. Acarol. Res.* **2019**, *51*, 8046. [CrossRef]
74. Liu, Z.; Morel, P.C.H.; Minor, M.A. Substrate and moisture content effects on pupation of the black soldier fly (Diptera: Stratiomyidae). *J. Insects Food Feed* **2022**, *9*, 415–425. [CrossRef]
75. Tegtmeier, D.; Hurka, S.; Klüber, P.; Brinkrolf, K.; Heise, P.; Vilcinskas, A. Cottonseed press cake as a potential diet for industrially farmed black soldier fly larvae triggers adaptations of their bacterial and fungal gut microbiota. *Front. Microbiol.* **2021**, *12*, 634503. [CrossRef]
76. Cammack, J.A.; Tomberlin, J.K. The impact of diet protein and carbohydrate on select life-history traits of the black soldier fly *Hermetia illucens* (L.) (Diptera: Stratiomyidae). *Insects* **2017**, *8*, 56. [CrossRef]
77. Lardé, G. Investigation on some factors affecting larval growth in a coffee-pulp bed. *Biol. Wastes* **1989**, *30*, 11–19. [CrossRef]
78. Palma, L.; Fernandez-Bayo, J.; Niemeier, D.; Pitesky, M.; VanderGheynst, J.S. Managing high fiber food waste for the cultivation of black soldier fly larvae. *NPJ Sci. Food* **2019**, *3*, 15. [CrossRef]
79. Tschirner, M.; Simon, A. Influence of different growing substrates and processing on the nutrient composition of black soldier fly larvae destined for animal feed. *J. Insects Food Feed* **2015**, *1*, 249–259. [CrossRef]
80. Yakti, W.; Müller, M.; Klost, M.; Mewis, I.; Dannehl, D.; Ulrichs, C. Physical properties of substrates as a driver for *Hermetia illucens* (L.) (Diptera: Stratiomyidae) larvae growth. *Insects* **2023**, *14*, 266. [CrossRef] [PubMed]

81. Bonelli, M.; Bruno, D.; Brilli, M.; Gianfranceschi, N.; Tian, L.; Tettamanti, G.; Caccia, S.; Casartelli, M. Black soldier fly larvae adapt to different food substrates through morphological and functional responses of the midgut. *Int. J. Mol. Sci.* **2020**, *21*, 4955. [CrossRef] [PubMed]
82. Alyokhin, A.; Buzza, A.; Beaulieu, J. Effects of food substrates and moxidectin on development of black soldier fly, *Hermetia illucens*. *J. Appl. Entomol.* **2019**, *143*, 137–143. [CrossRef]
83. Palma, L.; Ceballos, S.J.; Johnson, P.C.; Niemeier, D.; Pitesky, M.; VanderGheynst, J.S. Cultivation of black soldier fly larvae on almond byproducts: Impacts of aeration and moisture on larvae growth and composition. *J. Sci. Food Agric.* **2018**, *98*, 5893–5900. [CrossRef] [PubMed]
84. Shumo, M.; Osuga, I.M.; Khamis, F.M.; Tanga, C.M.; Fiaboe, K.K.M.; Subramanian, S.; Ekesi, S.; Van Huis, A.; Borgemeister, C. The nutritive value of black soldier fly larvae reared on common organic waste streams in Kenya. *Sci. Rep.* **2019**, *9*, 10110. [CrossRef] [PubMed]
85. Erickson, M.C.; Islam, M.; Sheppard, C.; Liao, J.; Doyle, M.P. Reduction of *Escherichia coli* O157:H7 and *Salmonella enterica* Serovar Enteritidis in chicken manure by larvae of the black soldier fly. *J. Food Prot.* **2004**, *67*, 685–690. [CrossRef]
86. Harnden, L.M.; Tomberlin, J.K. Effects of temperature and diet on black soldier fly, *Hermetia illucens* (L.) (Diptera: Stratiomyidae), development. *Forensic Sci. Int.* **2016**, *266*, 109–116. [CrossRef]
87. Barragan-Fonseca, K.B.; Dicke, M.; Van Loon, J.J.A. Influence of larval density and dietary nutrient concentration on performance, body protein, and fat contents of black soldier fly larvae (*Hermetia illucens*). *Entomol. Exp. Appl.* **2018**, *166*, 761–770. [CrossRef]
88. Chia, S.Y.; Tanga, C.M.; Khamis, F.M.; Mohamed, S.A.; Salifu, D.; Sevgan, S.; Fiaboe, K.K.M.; Niassy, S.; Van Loon, J.J.A.; Dicke, M.; et al. Threshold temperatures and thermal requirements of black soldier fly *Hermetia illucens*: Implications for mass production. *PLoS ONE* **2018**, *13*, e0206097. [CrossRef]
89. Klammsteiner, T.; Walter, A.; Bogataj, T.; Heussler, C.D.; Stres, B.; Steiner, F.M.; Schlick-Steiner, B.C.; Insam, H. Impact of processed food (canteen and oil wastes) on the development of black soldier fly (*Hermetia illucens*) larvae and their gut microbiome functions. *Front. Microbiol.* **2021**, *12*, 619112. [CrossRef] [PubMed]
90. Cheng, J.Y.K.; Chiu, S.L.H.C.; Lo, I.M.C. Effects of moisture content of food waste on residue separation, larval growth and larval survival in black soldier fly bioconversion. *Waste Manag.* **2017**, *67*, 315–323. [CrossRef] [PubMed]
91. Hosseindoust, A.; Ha, S.H.; Mun, J.Y.; Kim, J.S. Quality characteristics of black soldier flies produced by different substrates. *Insects* **2023**, *14*, 500. [CrossRef] [PubMed]
92. Grossule, V.; Vanini, S.; Lavagnolo, M.C. Potential treatment of leachate by *Hermetia illucens* (Diptera, Stratyomyidae) larvae: Performance under different feeding conditions. *Waste Manag. Res.* **2020**, *38*, 537–545. [CrossRef] [PubMed]
93. Ma, J.; Lei, Y.; Rehman, K.U.; Yu, Z.; Zhang, J.; Li, W.; Li, Q.; Tomberlin, J.K.; Zheng, L. Dynamic effects of initial pH of substrate on biological growth and metamorphosis of black soldier fly (Diptera: Stratiomyidae). *Environ. Entomol.* **2018**, *47*, 159–165. [CrossRef]
94. Meneguz, M.; Gasco, L.; Tomberlin, J.K. Impact of pH and feeding system on black soldier fly (*Hermetia illucens*, L.; Diptera: Stratiomyidae) larval development. *PLoS ONE* **2018**, *13*, e0202591. [CrossRef]
95. Lardé, G. Recycling of coffee pulp by *Hermetia illucens* (Diptera: Stratiomyidae) larvae. *Biol. Wastes* **1990**, *33*, 307–310. [CrossRef]
96. Li, Q.; Zheng, L.; Qiu, N.; Cai, H.; Tomberlin, J.K.; Yu, Z. Bioconversion of dairy manure by black soldier fly (Diptera: Stratiomyidae) for biodiesel and sugar production. *Waste Manag.* **2011**, *31*, 1316–1320. [CrossRef]
97. Klammsteiner, T.; Walter, A.; Bogataj, T.; Heussler, C.D.; Stres, B.; Steiner, F.M.; Schlick-Steiner, B.C.; Arthofer, W.; Insam, H. The core gut microbiome of black soldier fly (*Hermetia illucens*) larvae raised on low-bioburden diets. *Front. Microbiol.* **2020**, *11*, 993. [CrossRef]
98. Parodi, A.; Van Dijk, K.; Van Loon, J.J.A.; De Boer, I.J.M.; Van Schelt, J.; Van Zanten, H.H.E. Black soldier fly larvae show a stronger preference for manure than for a mass-rearing diet. *J. Appl. Entomol.* **2020**, *144*, 560–565. [CrossRef]
99. Julita, U.; Suryani, Y.; Kinasih, I.; Yuliawati, A.; Cahyanto, T.; Maryeti, Y.; Permana, A.D.; Fitri, L.L. Growth performance and nutritional composition of black soldier fly, *Hermetia illucens* (L), (Diptera: Stratiomyidae) reared on horse and sheep manure. In Proceedings of the IOP Conference Series: Earth and Environmental Science, Bogor, Indonesia, 19–20 October 2017; IOP Publishing: Bristol, UK, 2018; Volume 187. [CrossRef]
100. Diener, S.; Zurbrügg, C.; Tockner, K. Conversion of organic material by black soldier fly larvae: Establishing optimal feeding rates. *Waste Manag. Res.* **2009**, *27*, 603–610. [CrossRef] [PubMed]
101. Spranghers, T.; Ottoboni, M.; Klootwijk, C.; Ovyn, A.; Deboosere, S.; De Meulenaer, B.; Michiels, J.; Eeckhout, M.; De Clercq, P.; De Smet, S. Nutritional composition of black soldier fly (*Hermetia illucens*) prepupae reared on different organic waste substrates. *J. Sci. Food Agric.* **2017**, *97*, 2594–2600. [CrossRef] [PubMed]
102. Maurer, V.; Holinger, M.; Amsler, Z.; Früth, B.; Wohlfahrt, J.; Stamer, A.; Leiber, F. Replacement of soybean cake by *Hermetia illucens* meal in diets for layers. *J. Insects Food Feed* **2016**, *2*, 83–90. [CrossRef]
103. Miranda, C.D.; Cammack, J.A.; Tomberlin, J.K. Interspecific competition between the house fly, *Musca domestica* L. (Diptera: Muscidae) and black soldier fly, *Hermetia illucens* (L.) (Diptera: Stratiomyidae) when reared on poultry manure. *Insects* **2019**, *10*, 440. [CrossRef] [PubMed]
104. Veldkamp, T.; Van Rozen, K.; Elissen, H.; Van Wikselaar, P.; Van der Weide, R. Bioconversion of digestate, pig manure and vegetal residue-based waste operated by black soldier fly larvae, *Hermetia illucens* L. (Diptera: Stratiomyidae). *Animals* **2021**, *11*, 3082. [CrossRef] [PubMed]

105. Chia, S.Y.; Tanga, C.M.; Osuga, I.M.; Mohamed, S.A.; Khamis, F.M.; Salifu, D.; Sevgan, S.; Fiaboe, K.K.M.; Niassy, S.; Van Loon, J.J.A.; et al. Effects of waste stream combinations from brewing industry on performance of black soldier fly, *Hermetia illucens* (Diptera: Stratiomyidae). *PeerJ* **2018**, *6*, e5885. [CrossRef]
106. Oonincx, D.G.A.B.; Volk, N.; Diehl, J.J.E.; Van Loon, J.J.A.; Belušič, G. Photoreceptor spectral sensitivity of the compound eyes of black soldier fly (*Hermetia illucens*) informing the design of LED-based illumination to enhance indoor reproduction. *J. Insect Physiol.* **2016**, *95*, 133–139. [CrossRef] [PubMed]
107. Bertinetti, C.; Samayoa, A.C.; Hwang, S. Effects of feeding adults of *Hermetia illucens* (Diptera: Stratiomyidae) on longevity, oviposition, and egg hatchability: Insights into optimizing egg production. *J. Insect Sci.* **2019**, *19*, 19. [CrossRef]
108. Klüber, P.; Arous, E.; Zorn, H.; Rühl, M. Protein- and carbohydrate-rich supplements in feeding adult black soldier flies (*Hermetia illucens*) affect life history traits and egg productivity. *Life* **2023**, *13*, 355. [CrossRef]
109. Yakti, W.; Schulz, S.; Marten, V.; Mewis, I.; Padmanabha, M.; Hempel, A.-J.; Kobelski, A.; Streif, S.; Ulrichs, C. The effect of rearing scale and density on the growth and nutrient composition of *Hermetia illucens* (L.) (Diptera: Stratiomyidae) larvae. *Sustainability* **2022**, *14*, 1772. [CrossRef]
110. Park, K.; Kim, W.; Kim, E.; Choi, J.-Y.; Kim, S.-H. Effect of adult population density on egg production in the black soldier fly, *Hermetia illucens* (Diptera: Stratiomyidae). *Int. J. Ind. Entomol.* **2016**, *33*, 92–95. [CrossRef]
111. Tomberlin, J.K.; Sheppard, D.C. Factors influencing mating and oviposition of black soldier fly (Diptera: Stratiomyidae) in a colony. *J. Entomol. Sci.* **2002**, *37*, 345–352. [CrossRef]
112. Liu, Z.; Najar-Rodriguez, A.J.; Morel, P.C.H.; Minor, M.A. Reproduction of black soldier fly (Diptera: Stratiomyidae) under different adult densities and light regimes. *J. Econ. Entomol.* **2022**, *115*, 37–45. [CrossRef] [PubMed]
113. El-Dakar, M.A.; Ramzy, R.R.; Plath, M.; Ji, H. Evaluating the impact of bird manure vs. mammal manure on *Hermetia illucens* larvae. *J. Clean. Prod.* **2021**, *278*, 123570. [CrossRef]
114. Grausa, K.; Siddiqui, S.A.; Lameyer, N.; Wiesotzki, K.; Smetana, S.; Pentjuss, A. Metabolic modeling of *Hermetia illucens* larvae resource allocation for high-value fatty acid production. *Metabolites* **2023**, *13*, 724. [CrossRef] [PubMed]
115. Čičková, H.; Pastor, B.; Kozánek, M.; Martínez-Sánchez, A.; Rojo, S.; Takáč, P. Biodegradation of pig manure by the housefly, *Musca domestica*: A viable ecological strategy for pig manure management. *PLoS ONE* **2012**, *7*. e32798. [CrossRef] [PubMed]
116. James, M.T. The genus *Hermetia* in the United States (Diptera: Stratiomyidae). *Bull. Brooklyn Entomol. Soc.* **1935**, *30*, 165–170.
117. Holmes, L.A.; Vanlaerhoven, S.L.; Tomberlin, J.K. Lower temperature threshold of black soldier fly (Diptera: Stratiomyidae) development. *J. Insects Food Feed* **2016**, *2*, 255–262. [CrossRef]
118. Holmes, L.A.; Vanlaerhoven, S.L.; Tomberlin, J.K. Relative humidity effects on the life history of *Hermetia illucens* (Diptera: Stratiomyidae). *Environ. Entomol.* **2012**, *41*, 971–978. [CrossRef]
119. Heussler, C.D.; Walter, A.; Oberkofler, H.; Insam, H.; Arthofer, W.; Schlick-Steiner, B.C.; Steiner, F.M. Influence of three artificial light sources on oviposition and half-life of the black soldier fly, *Hermetia illucens* (Diptera: Stratiomyidae): Improving small-scale indoor rearing. *PLoS ONE* **2018**, *13*, e0197896. [CrossRef]
120. Klüber, P.; Bakonyi, D.; Zorn, H.; Rühl, M. Does light color temperature influence aspects of oviposition by the black soldier fly (Diptera: Stratiomyidae)? *J. Econ. Entomol.* **2020**, *113*, 2549–2552. [CrossRef] [PubMed]
121. Zhang, J.; Huang, L.; He, J.; Tomberlin, J.K.; Li, J.; Lei, C.; Sun, M.; Liu, Z.; Yu, Z. An artificial light source influences mating and oviposition of black soldier flies, *Hermetia illucens*. *J. Insect Sci.* **2010**, *10*, 202. [CrossRef] [PubMed]
122. Liu, Z.; Najar-Rodriguez, A.J.; Minor, M.A.; Hedderley, D.I.; Morel, P.C.H. Mating success of the black soldier fly, *Hermetia illucens* (Diptera: Stratiomyidae), under four artificial light sources. *J. Photochem. Photobiol. B Biol.* **2020**, *205*, 111815. [CrossRef] [PubMed]
123. Schneider, J.C. Effects of light intensity on mating of the black soldier fly (*Hermetia illucens*, Diptera: Stratiomyidae). *J. Insects Food Feed* **2019**, *6*, 111–119. [CrossRef]
124. Gerber, P.J. Tackling Climate Change through Livestock: A Global Assessment of Emissions and Mitigation Opportunities. Food and Agriculture Organization of the United Nations (FAO), 2013. Available online: http //www.fao.org/3/a-i3437e.pdf (accessed on 25 January 2023).
125. Bortolini, S.; Macavei, L.I.; Saadoun, J.H.; Foca, G.; Ulrici, A.; Bernini, F.; Malferrari, D.; Setti, L.; Ronga, D.; Maistrello, L. *Hermetia illucens* (L.) larvae as chicken manure management tool for circular economy. *J. Clean. Prod.* **2020**, *262*, 121289. [CrossRef]
126. Mazza, L.; Xiao, X.; Ur Rehman, K.; Cai, M.; Zhang, D.; Fasulo, S.; Tomberlin, J.K.; Zheng, L.; Soomro, A.A.; Yu, Z.; et al. Management of chicken manure using black soldier fly (Diptera: Stratiomyidae) larvae assisted by companion bacteria. *Waste Manag.* **2020**, *102*, 312–318. [CrossRef]
127. Liu, Q.; Tomberlin, J.K.; Brady, J.A.; Sanford, M.R.; Yu, Z. Black soldier fly (Diptera: Stratiomyidae) larvae reduce *Escherichia coli* in dairy manure. *Environ. Entomol.* **2008**, *37*, 1525–1530. [CrossRef]
128. Lopes, I.G.; Lalander, C.; Vidotti, R.M.; Vinnerås, B. Reduction of bacteria in relation to feeding regimes when treating aquaculture waste in fly larvae composting. *Front. Microbiol.* **2020**, *11*, 1616. [CrossRef]
129. Wynants, E.; Frooninckx, L.; Crauwels, S.; Verreth, C.; De Smet, J.; Sandrock, C.; Wohlfahrt, J.; Van Schelt, J.; Depraetere, S.; Lievens, B.; et al. Assessing the microbiota of black soldier fly larvae (*Hermetia illucens*) reared on organic waste streams on four different locations at laboratory and large scale. *Microb. Ecol.* **2019**, *77*, 913–930. [CrossRef]
130. Ur Rehman, K.; Rehman, A.; Cai, M.; Zheng, L.; Xiao, X.; Somroo, A.A.; Wang, H.; Li, W.; Yu, Z.; Zhang, J. Conversion of mixtures of dairy manure and soybean curd residue by black soldier fly larvae (*Hermetia illucens* L.). *J. Clean. Prod.* **2017**, *154*, 366–373. [CrossRef]

131. Bosch, G.; Oonincx, D.G.A.B.; Jordan, H.R.; Zhang, J.; Van Loon, J.J.A.; Van Huis, A.; Tomberlin, J.K. Standardisation of quantitative resource conversion studies with black soldier fly larvae. *J. Insects Food Feed* **2020**, *6*, 95–109. [CrossRef]
132. Broeckx, L.; Frooninckx, L.; Slegers, L.; Berrens, S.; Noyens, I.; Goossens, S.; Verheyen, G.; Wuyts, A.; Van Miert, S. Growth of black soldier fly larvae reared on organic side-streams. *Sustainability* **2021**, *13*, 12953. [CrossRef]
133. Romano, N.; Fischer, H.; Kumar, V.; Francis, S.A.; Sinha, A.K. Productivity, conversion ability, and biochemical composition of black soldier fly (*Hermetia illucens*) larvae fed with sweet potato, spent coffee or dough. *Int. J. Trop. Insect Sci.* **2022**, *42*, 183–190. [CrossRef]
134. Eggink, K.M.; Lund, I.; Pedersen, P.B.; Hansen, B.W.; Dalsgaard, J. Biowaste and by-products as rearing substrates for black soldier fly (*Hermetia illucens*) larvae: Effects on larval body composition and performance. *PLoS ONE* **2022**, *17*, e0275213. [CrossRef]
135. Scala, A.; Cammack, J.A.; Salvia, R.; Scieuzo, C.; Franco, A.; Bufo, S.A.; Tomberlin, J.K.; Falabella, P. Rearing substrate impacts growth and macronutrient composition of *Hermetia illucens* (L.) (Diptera: Stratiomyidae) larvae produced at an industrial scale. *Sci. Rep.* **2020**, *10*, 19448. [CrossRef]

agriculture

Article

Inclusion of Dried Black Soldier Fly Larvae in Free-Range Laying Hen Diets: Effects on Production Efficiency, Feed Safety, Blood Metabolites, and Hen Health

Masoumeh Bejaei [1,2,*] and Kimberly M. Cheng [1]

1 Avian Research Centre, Faculty of Land and Food Systems, University of British Columbia, 2357 Main Mall, Vancouver, BC V6T 1Z4, Canada; kimberly.cheng@ubc.ca
2 Summerland Research and Development Centre, Science and Technology Branch, Agriculture and Agri-Food Canada, Summerland, BC V0H 1Z0, Canada
* Correspondence: masoumeh.bejaei@agr.gc.ca

Abstract: Identifying alternative feedstuffs to replace conventional nutrient sources in poultry diets is crucial to supplying the growing demand for animal feed. A 17-week-long feeding experiment with three diets, including 0% (control), 10%, and 18% full-fat dried black soldier fly larvae (DBSFL), was conducted to evaluate the production efficiency and feed safety of using the larvae for partial (50%) and full (100%) substitutions of soybean meal and 90% replacement of soybean oil in free-range laying hen diets. Thirty hens (18–36 weeks old) were housed in two mobile poultry trailers per treatment level. The weight gain of hens, their feed intake, egg production, egg weights, feed conversion ratios, bird welfare parameters, hematology and blood metabolites, fecal microbiology, and digestive tract weights were examined. Control hens had higher weight gains, laid more and bigger eggs while consuming less feed, and had lower feed conversion ratios than 18% DBSFL hens. However, the production performances of 10% DBSFL hens were not significantly different from the control in many of the parameters considered (e.g., hen-day egg production or HDEP). In conclusion, partial replacement of soybean meal and oil with DBSFL in layer diets maintains production efficiency, feed safety, and hen health and welfare status.

Keywords: laying hen; dried black soldier fly larvae; *Hermetia illucens*; edible insects; production efficiency; hen health; feed safety; egg production; digestive tract; blood metabolites

1. Introduction

Soybean meal and oil are excellent sources of protein and metabolizable energy (respectively) in laying hen diets, and they have been widely used as feed ingredients in commercial poultry production for several decades. The consumer demand for animal products (in particular poultry products) has shown significant growth in the last few decades. The latest projections indicate that global demand for food in general will increase at least by 50–60% from 2019 to 2050 [1]. This demonstrates that the need for animal feed will also increase. Soybean cultivation for poultry feed requires large areas of arable land that otherwise can be used to produce plant-based sources of food for direct human consumption. Thus, it is important to identify extra feed sources, using alternative nutrient sources that are not used directly for human consumption, to supply the growing demand for animal feed [2].

Animal processing by-products (e.g., meat meal and blood meal) and fish meal have also been used as sources of protein in poultry feed for decades [3]. However, the use of animal by-products is severely restricted in some countries (e.g., the European Union [4] because of public health concerns related to their use [5]). Furthermore, the use of animal by-products is not allowed in organic production systems (e.g., the National Standard of Canada [6], indicating that organic poultry farmers also face a scarcity of protein sources for organic poultry diets [7]. In addition, fish meal prices have increased significantly over time due to static production and a growing demand from the aquaculture industry [8].

The need for alternative feed sources in animal production has facilitated the development of circular food systems in which food waste, by-products, and inedible nutrients for humans can be used to produce feed for animals [2]. When the reduction of food loss and waste is not feasible, reusing the lost and wasted nutrients by recycling them through circular food systems is of great importance. Nevertheless, there are diverse standards and guidelines in different regions concerning which lost or wasted food products can be recycled back directly into the food system through animal feed. About one-third (31%) of the food produced in the world is lost or wasted, and approximately half of this wastage (14%) happens at the retail and consumer levels [8]. At this stage, the nutrients are still of high quality and not contaminated. Insects and worms have shown potential for recovering nutrients from food waste by establishing a natural recycling and recovery system for the production of highly nutritious feedstuffs. Their production is considered a viable solution for tackling both the increased demand for animal feed and the challenges caused by the production of excessive amounts of food waste in our current food systems [2,3,9,10].

Many alternative nutrient sources have been tested in the last few decades to identify novel feed ingredients that can improve the environmental, economic, and social sustainability of animal farming while increasing feed production. For example, the use of insects in aquaculture, poultry, and swine production has been authorized in several countries/regions [11]. Nevertheless, there are considerable differences in the insect production systems and conditions in different regions (e.g., what is allowed to be used as the nutrient source for insects). These differences have made comparison of published results difficult (see discussion) [9,10,12].

The black soldier fly (*Hermetia illucens*) is one of the insects used to convert significant amounts of organic waste into biomass rich in protein and lipid [3,13,14]. The use of black soldier fly larvae (from now on referred to as BSFL) has been promising in aquaculture and monogastric animal production as a substitute for conventional lipid and protein sources in aquaculture and poultry feed [2,10,14–22]. The nutritional value of BSFL as a feedstuff depends on the nutrient sources provided to the insect and the production stage of BSFL [21,23]. Different studies have used BSFL in different formats (full fat, defatted, chopped, grounded, meal, etc.) at different stages of animal production. Moreover, potential pathogen contents of the larvae (e.g., Escherichia coli O157:H7 and Salmonella spp. and Enterococcus spp.) and the presence of heavy metals (such as As, Cd, Cr, and Pb) or other chemical contamination are concerns about their use as a feed ingredient [24–26], but investigations have rarely been reported in the published literature.

Considering the knowledge gaps in this area, this study was conducted to investigate the effects of the inclusion of full-fat dried BSFL (from here on referred to as DBSFL), either as partial or full replacements of soybean meal and oil, on the production efficiency (i.e., weight gain of hens, their feed intake, egg production, egg weight and feed conversion ratio), bird health and welfare parameters, hematology and blood metabolites, fecal microbiology and digestive tract length, and weight of commercial laying hens, from the beginning of their egg production until their peak egg production. The DBSFL used in the current study was produced from pre-consumption plant-based food waste.

2. Materials and Methods

The current experiment was conducted under the guidelines of the Canadian Council on Animal Care, and the study was certified by the University of British Columbia (UBC) Animal Care Committee (Certificate #A16-0227).

2.1. Source of DBSFL

Whole DBSFL provided by Enterra Feed Corporation (Langley, BC, Canada) was used in this study as a protein and lipid source in the laying hen diets in the form of chopped larvae (about 3–5 mm length). The larvae were produced from pre-consumption plant-based food waste that was mainly supplied by the food processing and retail sectors. The larvae were harvested as a 4th or 5th instar prior to darkening into prepupae to minimize chitin content. Further information about the production practices was considered proprietary and was not shared by the producer.

2.2. Experimental Design

The feed trial was conducted at the Kwantlen Polytechnic University Teaching and Education Centre (Richmond, BC, Canada), and the laboratory work was conducted at the UBC Avian Research Centre (Vancouver, BC, Canada).

A Completely Randomized Design (CRD) study with 3 dietary treatment levels (0%, 10%, and 18% DBSFL) and 2 replications per diet was carried out for 17 weeks, starting when the hens were 18 weeks old.

Ninety commercial brown layer pullets (Novogen BROWN) were obtained from a local supplier. Each experimental unit, consisting of 15 layers, was housed in a 3.35 m (L) × 1.5 m (W) × 1.8 m (H) rat and predator-proof mobile poultry trailer, custom-built according to Canadian General Standards Board guidelines for Organic production systems [6]. The outside free-range area provided 0.33 m^2 per hen, and the off-the-ground area inside the hut area provided 0.17 m^2 per hen. Each of the six trailers was equipped with four 35.7 cm (L) × 33.0 cm (W) × 35.7 cm (H) nest boxes, inside perches (~20 cm/hen), two feeders (one inside and one outside under cover), a 13.25 liter water bucket with four nipple drinkers, and a sandbox for dust bathing. Trailers provided an enclosed environment for hens by covering all areas with wire mesh except for the ones made of wood (Figure S1). All six trailers were located in a field seeded with rye grass and white clover. Trailers were moved to new ground once every two weeks so that all birds could have access to fresh cover crops at the same time. A solar-powered electric fence was installed all around the field to minimize human and animal disturbances. From here on, the three dietary treatment groups will be referred to as 0% DBSFL (or control), 10% DBSFL, and 18% DBSFL, respectively.

2.3. Diet Formulation and Feed Nutrient Analysis

Before formulating the diets (Table 1), the proximate compositions of the DBSFL used in this study were analyzed. Proximate composition analysis indicated that its moisture, dry matter, crude protein, crude fiber, crude fat, and crude ash contents were 4.2%, 95.8%, 37.9%, 7.4%, 43.6%, and 6.4%, respectively. The ME value of DBSFL was calculated at 5.2 kcal/g based on the Carpenter and Clegg equation ($-53 + 38$ (% crude protein + 2.25 × % ether extract + 1.1 × % starch + % sugar)) [27]. This information was used to formulate the three experimental diets using the WUFFDA program (University of Georgia Extension, Athens, GA, 2014). The diets were formulated to be isocaloric and isonitrogenous according to NRC [28] laying hen nutrition requirements. In the partial replacement diet, 50% of the soybean meal was removed and 10% DBSFL could be included, along with some modifications in the inclusion rates of other feedstuffs to balance the diet to the same protein and energy levels as the control diet. In the full replacement diet, all the soybean meal was removed, and 18% DBSFL could be included with some modifications in the inclusion rates of other ingredients to balance the diet to the same energy and protein levels. All three diets met the requirements for essential amino acids and fatty acids and were prepared by a local organic feed supplier (In Season Farms, Abbotsford, BC). Feed was provided in the mash form in all three diets, and whole DBSFL were chopped (size of 3–5 mm) before their inclusion in the treatment diets.

In the 10% DBSFL and 18% DBSFL diets, DBSFL substituted 50% and 100% of the soybean meal, respectively. Moreover, in both diets, DBSFL also replaced 90% of the soybean oil because larvae have a high lipid content. The chemical composition, fatty acid profile, 12 amino acid concentrations, and metal profile of the three experimental diets were analyzed based on the standards defined by AOAC [29].

Table 1. Ingredients and calculated composition of three experimental diets [1].

Ingredients (% *w*/*w*)	Diets		
	0% DBSFL [2]	10% DBSFL	18% DBSFL
Chopped DBSFL	0	10	18
Soybean meal	20	10	0
Sunflower meal	14.3	12.3	15
Soybean oil	4.95	0.5	0.5
Ground Corn	49.5	41.5	30
Barley	0	3.16	10
Wheat	0	9	4
Wheat bran	0	1	10
Dehydrated alfalfa meal	1	3	3.4
DL-Methionine	0.15	0	0
Limestone	8.65	8.12	7.7
Dicalcium phosphate	0.5	0.5	0.5
Common Salt	0.35	0.32	0.3
Vitamin Mineral Premix	0.4	0.4	0.4
Choline	0.1	0.1	0.1
Phytase Enzyme	0.1	0.1	0.1
Calculated composition			
Dry matter (%)	91.13	91.24	91.83
M.E. (Kcal/g)	2.79	2.79	2.79
Crude protein (%)	17.02	17.02	17.02
Ether extract (%)	7.9	6.73	9.9
Calcium (%)	3.52	3.53	3.53
Na (%)	0.18	0.18	0.18
Total phosphorus (%)	0.54	0.51	0.59

[1] Feedstuff nutrient values provided by NRC [28] and Feed Supplier (In Season Farms), and laying hen nutrient requirements provided by [28,30] considering 120 g feed/hen/day; 2 DBSFL: Dried black soldier fly larvae.

2.4. Experimental Protocols and Production Performance Data

Pullets arrived as 18-week-old birds in the last week of June. The first two weeks were an adaptation period, and the birds were fed the control diet ad libitum. All birds were weighed and tagged with numbered wing tags at the end of the first week. After the second week, all the birds were gradually introduced to their respective diet treatments. By the end of the third week, the transition and adaptation stages were completed. The data collection started at the beginning of Week 4, when the birds were 22 weeks old, and continued until Week 17 (Week 36). This study focused on the early egg production stage because it can have a long-lasting effect on production efficiency and hen health. During this stage, hens not only initiate laying eggs, reach their highest laying rate, and complete a 10-week peak production, but also gain considerable body weight [31]. Individual hen weights were recorded biweekly during the experiment period (total of 7 times).

Each trailer was provided daily with 2.2 kg of feed evenly distributed between the two feeders, and the leftover feed was weighed weekly using a handheld digital scale (Salter-Brecknell SA3N253, ElectroSamson Digital Scale, Fairmont, MN). Then, the daily feed intake per hen and feed conversion ratio (FCR) were calculated.

The number of eggs produced in each trailer was recorded daily, and then the weekly hen-day egg production (HDEP) was calculated.

Daily egg production in each trailer was weighed once a week using a pocket scale (100 g with 0.01 g reliability), and the results were extrapolated to represent the mean weekly egg weight per trailer.

2.5. Blood Chemistry and White Blood Cells (WBC) Differential Counts

Baseline blood samples (1 mL) were collected from three randomly selected hens in each trailer from the brachial vein using a tuberculin syringe with a 25 G heparinized needle on Week 2 before the hens were put on the experimental diets. The second blood sample was collected in Week 17 from the same birds immediately after decapitation (for post-mortem examination) by collecting trunk blood into heparinized vacutainer tubes. Blood was stored in BD Vacutainer® Heparin Tubes (light green-topped tubes, Franklin Lakes, NJ, USA) and transported to the laboratory immediately after.

Blood biochemistry metabolites were analyzed to determine the concentrations of albumin, albumin/globulin ratio, amylase, aspartate transaminase enzyme, bile acids, blood urea nitrogen, calcium, chloride, cholesterol, creatine kinase, creatinine, gamma-glutamyl transferase enzyme, globulin, glucose, phosphorus, potassium, sodium, sodium/potassium ratio, total protein, triglycerides, and uric acid. Blood smears were also made for the WBC differential counts. Heterophil: Lymphocyte (H:L) ratios were calculated by dividing the absolute heterophil count by absolute lymphocyte counts. Moreover, packed cell volume (PCV or hematocrit) was measured after centrifuging blood in microhematocrit tubes at 12,000× *g* for 5 min.

2.6. Fecal Microbial Analysis

Fecal culture examination was conducted at Week 16, close to the end of the 17-week experiment, to study the pathogen content of hens and monitor the impact of the inclusion of the novel feed ingredient on hen health. This was carried out to ensure that the hens were exposed to the new ingredient for the longest time possible in this experiment and to detect any potential influence of the use of DBSFL on hen pathogen contents. About 5 g of fresh fecal samples were collected from each trailer by placing five randomly selected birds from each trailer in a clean crate with a plastic bag under the crate for half an hour. For each trailer, three fecal samples were pooled in a ProtocolTM C&S Medium container (Cary and Blair Medium containing 15 mL of Phenol Red solution; Fishe Diagnostics, VA, USA) (in total, six samples) for the detection of enteric pathogens. Standard bacterial growth media were used for specific incubation times and temperatures determined by the standard procedures applied by IDEXX Laboratory (Delta, BC, Canada) for the detection of enteric pathogens.

2.7. Health and Welfare Parameters of Laying Hens

Feather and wound scores [32] were recorded biweekly when the hens were weighed. Plumage conditions in six body parts (neck, breast, cloaca/vent, back, wings, and tail), bumble foot lesions, and pecking damage to the comb and the skin of the rear body were assigned scores of 1 to 4 (4 being the best integument conditions and 1 being the worst) [32].

2.8. Post-Mortem Examination of the Gastrointestinal Tract

At the end of the feed trial (Week 17), the three hens per trailer (in total, 18 hens) that were sampled in the baseline sampling were transported to the Necropsy Laboratory of the UBC Centre for Comparative Medicine. A post-mortem examination was performed to assess the effects of the three diets on the gastrointestinal tract weight of laying hens. The digestive tract of the hens was segmented into the proventriculus, gizzard (ventriculus), small intestine (duodenum, jejunum, and ileum), ceca, large intestine, liver, and pancreas [33]. The segments were weighed using a precision electric balance (0.01 g reliability). The proportional weights of the digestive tract segments were calculated using the segment weight divided by the bird's live weight taken just before decapitation. The individual segment weights divided by the total intestinal tract weight were also calculated as the digestive tract weight proportions.

2.9. Statistical Analysis

CRD was considered the general design of the experiment, with two replications (trailers) and a single-fixed effect treatment (with three levels: 0%, 10%, and 18% DBSFL). Trailers were assigned randomly to the treatment levels.

The experiment was designed to benefit from the strengths of the repeated measures mixed models as discussed by Schank and Koehnle [34] and Olejnik and Algina [35]. The application of repeated measures tests makes it possible to obtain unbiased results with limited samples [34]. These tests increase the precision of measurements, result in a reduction of measurement errors, and increase the likelihood of identifying statistically significant differences [34,35].

The α-level was 0.05 in all tests. The data were analyzed by Least Squares ANOVA with repeated measures using JMP® PRO software (version 17.0.0, SAS Institute Inc., Cary, NC, USA). Tukey HSD post hoc tests were used for means separation depending on the frequency of measurements, the presence of subsamples, and the type of analysis. Trailers were considered the experimental units, and the statistical models listed below were used in the present study to determine differences between treatment groups.

- CRD with repeated subsampling was used when measurements were conducted repeatedly during the experiment on hens (i.e., subsamples) within each trainer (i.e., experimental unit), e.g., blood biochemistry, WBC differential, and hen weight.
- CRD repeated in time were used when measurements were conducted repeatedly on the trailers without any subsampling, e.g., feed intake, HDEP, egg weight, and FCR.

Thus, the power of the tests was calculated for the CRD and repeated in time to check if the sample size was adequate for the tests. Cohen [36] recommended 0.8 as the generally accepted minimum level of power for tests.

Due to the limited sample size, statistical analyses were not performed on the feed composition and post-mortem data. However, the data are presented because they can provide valuable insights and trends that can be used to inform future research.

3. Results

3.1. Feed Composition Analyses

Proximate composition of the control, 10% DBSFL, and 18% DBSFL diets indicated their moisture (9.3%, 10.2%, 9.8%), dry matter (90.7%, 89.8%, 90.2%), crude protein (20.1%, 18.5%, 19.9%), crude fiber (7.3%, 7.2%, 9.1%), and crude fat (9.4%, 8.1%, 12.6%) contents (respectively). The crude fat concentration was higher in the 18% DBSFL diet because its lipid content was higher in absolute terms compared to the control and 10% DBSFL diets.

3.1.1. Metal Profiles of Diets

Table 2 shows that similar mineral mixes were included in the three diets, in approximately similar concentrations. The Ca concentrations were higher than the recommended level in all three experimental diets used in this study. It was higher in the DBSFL diets compared to that of the control feed. The 18% DBSFL diet had the lowest concentrations of Cu and Cr compared to the other two experimental diets. Zn concentrations were similar in three diets. The As, Cd, and Pb concentrations of the three diets were negligible.

Table 2. Metal profiles of three experimental diets.

Metal (ppm on Dry Matter Basis)	Diets 0% DBSFL	10% DBSFL [1]	18% DBSFL
Aluminum	134.53	186.13	180.43
Antimony	<5.00	<5.00	<5.00
Arsenic	<2.50	<2.50	<2.50
Barium	3.97	4.78	6.51
Boron	11.48	7.56	5.96
Cadmium	<0.50	<0.50	<0.50
Calcium	41,087.08	48,163.13	44,186.61
Chromium	7.17	7.22	6.07
Cobalt	<1.00	<1.00	<1.00
Copper	25.49	26.34	20.29
Iron	269.29	300.03	261.39

Table 2. *Cont.*

Metal (ppm on Dry Matter Basis)	Diets		
	0% DBSFL	10% DBSFL [1]	18% DBSFL
Lead	<2.50	<2.50	<2.50
Magnesium	1941.29	2052.45	2546.60
Manganese	162.56	154.13	181.21
Mercury	<10.00	<10.00	<10.00
Molybdenum	0.99	1.00	0.88
Phosphorus	5439.80	5080.56	6159.70
Potassium	9695.40	8287.59	8569.54
Selenium	<10.00	<10.00	<10.00
Sodium	2285.62	1984.66	1701.78
Sulfur	2808.74	2450.27	2421.97
Thallium	<10.00	<10.00	<10.00
Zinc	156.16	160.13	161.46

[1] DBSFL: Dried black soldier fly larvae.

3.1.2. Amino Acid Profiles of Diets

Figure 1 shows the amino acid content of three experimental diets. Threonine, valine, isoleucine, and lysine concentrations were higher in the three experimental diets than the recommended concentrations by Novogen [30]. Methionine concentrations of DBSFL diets were slightly lower than those of the control diet and the recommended concentration in the laying hen diets [30].

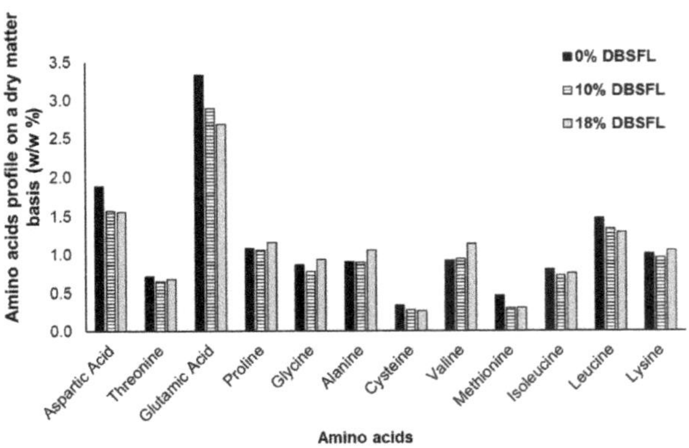

Figure 1. Amino acid analysis (*w*/*w*%, dry matter basis) of three experimental diets with different levels of dried black soldier fly larvae (DBSFL).

3.1.3. Fatty Acids Profiles of Diets

The results of the fatty acid analysis indicated that the lipid (ether extract) contents of the three diets seemed to differ from each other (Figure 2). The lipid content of the 0%, 10%, and 18% DBSFL diets was 8.48%, 7.27%, and 11.3% (as fed), or 9.35%, 8.1%, and 12.6% (dry matter), respectively. The lauric acid and myristic acid contents of both DBSFL diets were higher than those of the control feed. In addition to that, the palmitic acid and oleic acid contents of the 18% DBSFL diet were higher than those of the control group. On the other hand, linoleic acid and linolenic acid levels were higher in the control feed than those in the DBSFL diets.

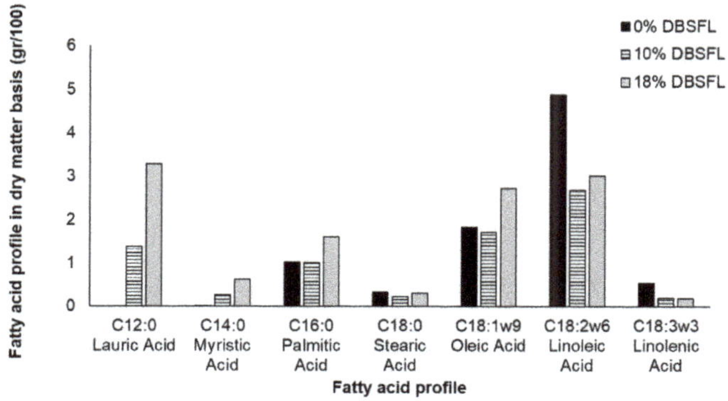

Figure 2. Fatty acid profiles of the three experimental diets with different levels of dried black soldier fly larvae (DBSFL).

3.2. Hen Weight

There was a significant ($p = 0.014$) Time × Diets interaction. The weight gains of hens from Week 4 to Week 16 differed depending on the level of DBSFL in the diet (Figure 3). Results indicated that the control hens gained more weight at a faster rate during the experiment compared to the weight gains of the hens in the 10% and 18% DBSFL diets. However, the live weight of 10% DBSFL hens was not significantly different from that of the 18% DBSFL hens throughout the trial, until the last weighing ($p < 0.05$).

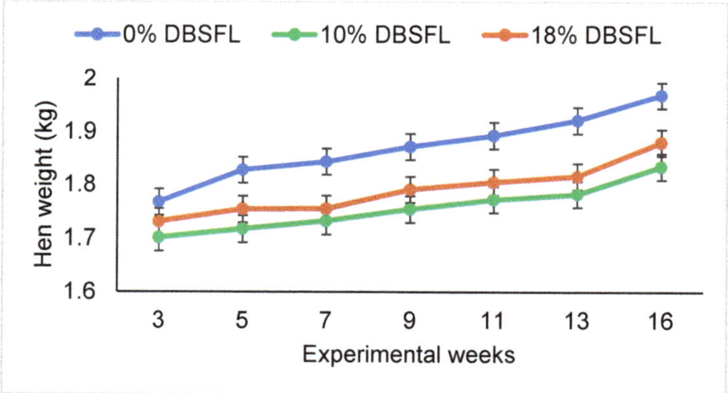

Figure 3. Time × Diet interaction effect on hen weights from Week 3 (age 21-week-old) until Week 16 of the experiment fed three experimental diets with different levels of dried black soldier fly larvae (DBSFL). Means followed by different letters were significantly different ($p < 0.05$); n = 42.

3.3. Feed Intake

There was a significant ($p = 0.04$) Time × Diet interaction effect (Table 3). Hens from the 18% DBSFL dietary treatment level consumed more feed compared to hens from the other two groups. The differences in the feed intake of the 18% DBSFL hens and the control hens were significant at Weeks 12, 13, and 16 ($p < 0.05$).

Table 3. Fixed effect tests and interaction effects Least Square Means (LS Means) of Time × Diet on the feed intake, hen-day egg production (HDEP), egg weight, and feed conversion ratio (FCR) from Week 4 (age 22-week-old) until Week 16 of the experiment fed three diets with different levels of dried black soldier fly larvae (DBSFL). (The means followed by different letters were significantly different at $p < 0.05$ and n = 78.)

		Feed Intake		HDEP		Egg Weight		FCR	
		F Ratio	*p*-Value (Power)	F Ratio	*p*-Value (Power)	F Ratio	*p*-Value (Power)	F Ratio	*p*-Value (Power)
Fixed Effect Tests	Treatment	12.48	0.04 (1.00)	1.75	0.31 (1.00)	73.99	<0.01 (1.00)	17.31	0.02 (1.00)
	Time	2.73	0.01 (0.94)	2.88	0.01 (0.95)	17.52	<0.0001 (1.00)	1.42	0.20 (0.81)
	Interaction	1.92	0.04 (0.93)	4.19	<0.0001 (1.00)	1.82	0.05 (0.91)	4.64	<0.0001 (1.00)
Diets	**Experi. weeks**	**LS Mean**	**Sig. level**	**LS Mean**	**Sig. level**	**LS Mean**	**Sig. level**	**LS Mean**	**Sig. level**
0% DBSFL	4	123.33	ABCDEF	0.98	ABCDE	56.89	HIJK	2.21	ABCDEFGH
	5	123.33	ABCDEF	0.97	ABCDE	59.52	CDEFGHIJ	2.13	CDEFGHI
	6	122.33	ABCDEF	0.97	ABCDE	60.70	ABCDEFG	2.07	DEFGHI
	7	120.33	CDEF	0.99	ABCD	62.28	ABCDEF	1.95	GHI
	8	121.67	BCDEF	0.97	ABCDE	62.50	ABCD	2.01	EFGHI
	9	121.33	BCDEF	0.98	ABCDE	63.05	AB	1.96	GHI
	10	118.67	EF	0.99	ABCDE	63.12	A	1.91	GHI
	11	120.00	CDEF	0.97	ABCDE	62.31	ABCDE	1.99	FGHI
	12	118.33	EF	0.97	ABCDE	62.58	ABC	1.96	GHI
	13	114.67	F	1.00	ABC	62.86	ABC	1.82	I
	14	126.00	ABCDEF	0.99	ABCD	63.16	A	2.01	EFGHI
	15	127.00	ABCDEF	0.98	ABCDE	62.55	ABC	2.07	DEFGHI
	16	119.33	DEF	1.00	ABC	63.22	A	1.89	HI
10% DBSFL	4	126.67	ABCDEF	0.98	ABCDE	55.43	K	2.34	ABCDEFGH
	5	125.33	ABCDEF	0.97	ABCDE	56.18	IJK	2.30	ABCDEFGHI
	6	124.00	ABCDEF	0.94	ABCDE	57.09	HIJK	2.30	ABCDEFGH
	7	127.67	ABCDEF	0.97	ABCDE	59.49	CDEFGHIJ	2.22	ABCDEFGHI
	8	134.00	ABCDEF	0.97	ABCDE	58.77	EFGHIJK	2.35	ABCDEFGH
	9	134.00	ABCDEF	0.98	ABCDE	59.92	ABCDEFGH	2.28	ABCDEFGHI
	10	127.00	ABCDEF	0.97	ABCDE	58.98	DEFGHIJ	2.23	ABCDEFGHI
	11	127.67	ABCDEF	0.99	ABCDE	60.17	ABCDEFGH	2.15	CDEFGHI
	12	124.67	ABCDEF	0.95	ABCDE	59.93	ABCDEFGH	2.19	BCDEFGHI
	13	127.33	ABCDEF	0.99	ABCDE	60.22	ABCDEFGH	2.15	CDEFGHI
	14	130.00	ABCDEF	0.97	ABCDE	59.50	BCDEFGHIJ	2.25	ABCDEFGHI
	15	131.67	ABCDEF	0.97	ABCDE	60.12	ABCDEFGH	2.25	ABCDEFGHI
	16	129.67	ABCDEF	0.97	ABCDE	60.38	ABCDEFGH	2.22	ABCDEFGHI
18% DBSFL	4	125.33	BCDEF	0.98	ABC	55.99	JK	2.28	BCDEFGHI
	5	131.67	ABCDEF	1.01	A	57.43	GHIJK	2.27	CDEFGHI
	6	129.67	ABCDEF	1.00	AB	57.32	GHIJK	2.27	CDEFGHI
	7	132.00	ABCDEF	0.97	ABCD	57.59	GHIJK	2.36	ABCDEFGH
	8	136.00	ABCDE	0.98	ABC	58.74	FGHIJK	2.36	ABCDEFG
	9	141.33	AB	0.98	ABCD	58.13	GHIJK	2.49	ABCD
	10	137.33	ABCDE	0.97	ABCDE	58.01	GHIJK	2.45	ABCDEF
	11	135.00	ABCDE	0.95	ABCDE	57.46	GHIJK	2.47	ABCDE
	12	142.33	A	0.91	DE	58.63	GHIJK	2.67	A
	13	139.00	ABCD	0.93	BCDE	58.02	GHIJK	2.58	ABC
	14	140.00	ABC	0.91	CDE	58.35	GHIJK	2.63	AB
	15	139.00	ABCD	0.94	BCDE	58.03	GHIJK	2.55	ABC
	16	142.33	A	0.90	E	59.51	BCDEFGHI	2.66	A
SE		3.38		0.02		0.60		0.08	

3.4. Hen-Day Egg Production (HDEP)

There was a significant ($p < 0.0001$) Time × Diet interaction effect on HDEP (Table 3). The results indicated that control and 10% DBSFL hens maintained high HDEP during the study (Weeks 4 to 16), but the HDEP of 18% DBSFL hens started to decrease at Week 11 and had lower HDEP than the control hens in the last 4 weeks of the study ($p < 0.05$), even though they still maintained above 90% HDEP.

3.5. Egg Weight

The mean egg weights from the 0%, 10%, and 18% DBSFL diets during the experiment were 61.90 ± 0.24 g, 58.93 ± 0.24 g, and 57.94 ± 0.24 g, respectively. The average weights of the eggs from all three diets were within the weight range for large eggs in the Canadian egg grading system (56 g to 63 g). There was a significant ($p < 0.05$) Time × Diet interaction (Table 3) in the egg weight variable. In Weeks 4 and 5 of the feed trial, eggs produced by hens in the three treatment groups were not significantly different in weight ($p < 0.05$). After Week 6, the egg weight of the 10% DBSFL diet started to show some inconsistent pattern in that they were significantly lighter than those from the control diet in Weeks 6, 8, 10, and 14 ($p < 0.05$) but not significantly different at other time points. The egg weights of the 18% DBSFL diet were significantly lighter than control eggs after Week 6 ($p < 0.05$). There were no significant differences in the egg weights of the 10% and 18% DBSFL diets throughout the study.

3.6. Feed Conversion Ratio (FCR)

Weekly FCR was significantly ($p < 0.0001$) affected by Time × Diet interaction (Table 3). Control hens had the best FCR throughout the study. Starting at Week 9, the FCR of 18% DBSFL hens was significantly higher than the FCR of the control hens ($p < 0.05$). Except for the first 3 weeks, the FCR of 10% DBSFL hens was between the FCRs of the control and the 18% DBSFL hens. The 10% DBSFL hens' FCR was significantly better than the 18% DBSFL hens at Week 12 ($p < 0.05$). Both 0% and 10% DBSFL hens maintained their FCR throughout the study, but the FCR of 18% DBSFL hens gradually increased.

3.7. Blood Metabolites

3.7.1. Blood Biochemistry

The baseline blood samples were taken during the ready-to-lay stage (20 weeks old), and as a result, changes in the blood parameters after 15 weeks of laying eggs were expected (Table 4), but for some of the blood parameters, these changes depended on Diet treatment levels (Table 5) or Time × Diet interactions (Table 6). The lipid content of the 18% DBSFL diet was higher than that of the control diet, but plasma triglycerides were higher in the control hens than in the 18% DBSFL hens (Table 5).

Table 4. There was a significant difference in plasma parameters between 20-week-old hen baseline plasma samples and 35-week-old hen blood samples (36 samples within 12 experimental units).

Plasma Parameters	Age		SEM	F Ratio	*p*-Value
	20 Weeks	35 Weeks			
Glucose (mmol/L)	4.27	9.73	0.46	27.73	0.01
Urea (Bun) (mmol/L)	0.32	0.68	0.04	20.69	0.01
Uric acid (mmol/L)	437.06	149.5	21.05	176.07	0.001
Calcium (mmol/L)	6.28	4.77	0.15	57.64	0.005
Sodium/Potassium ratio	39.50	26.17	0.67	640.00	0.0001
Chloride (mmol/L)	118.56	124.17	0.86	37.09	0.01
Total protein (g/L)	49.89	42.61	0.96	16.74	0.03
Albumin (g/L)	17.94	14.83	0.33	50.58	0.006
Aspartate Transaminase (IU/L)	141.5	200.56	4.61	134.86	0.001
Gamma GT (IU/L)	10.78	51.00	4.28	60.43	0.004
Cholesterol (mmol/L)	2.87	2.14	0.12	18.31	0.02
Triglycerides (mmol/L)	13.53	9.04	0.65	17.88	0.02
Bile acids (μmol/L)	36.87	6.91	1.72	961.11	<0.0001

Table 5. Significant Diet effect on plasma parameters (36 samples within 12 experimental units).

Plasma Parameters [1]	Diets				F Ratio	*p*-Value
	0% DBSFL [2]	10% DBSFL	18% DBSFL	SEM		
Total protein (g/L)	49.25a	46.67a	42.83b	0.60	28.68	0.01
Albumin (g/L)	17.92a	16.17ab	15.08b	0.44	10.77	0.04
Triglycerides (mmol/L)	13.05a	12.19ab	8.62b	0.64	13.40	0.03

[1] In each row, means followed by the same letter are not significantly different. [2] DBSFL: Dried black soldier fly larvae.

Table 6. Significant ($p < 0.05$) [1] Time × Diet interaction effect on plasma parameters (36 samples within 12 experimental units).

Plasma Parameters [1]	Diets								
	0% DBSFL [2]		10% DBSFL		18% DBSFL		SEM	F Ratio	*p*-Value
	Week 2 [3]	Week 17	Week 2	Week 17	Week 2	Week 17			
Creatinine (μmol/L)	14.33	15.00	14.83	11.33	12.17	8.67	1.31	12.50	0.04
Potassium (mmol/L)	3.77b	6.15a	3.85b	5.88a	3.90b	5.65a	0.15	1850	<0.0001
Amylase (IU/L)	387.50ab	385.50ab	422.33a	332.00ab	397.50a	277.50b	16.16	10.66	0.04
Creatine Kinase (IU/L)	613.33c	1904.00ab	721.17c	2336.33a	663.67c	1552.83b	64.19	20.52	0.02

[1] In each row, means followed by different letters were significantly different ($p < 0.05$); [2] DBSFL: Dried black soldier fly larvae; [3] At Week 2 of the experiment, hens were 20 weeks old, and hens in all three treatment groups were still on the control diet.

3.7.2. WBC Differential Counts

Most of the WBC differential counts were significantly affected by Time (Table 7), and, except for monocytes, the effect of the diet and the Diet × Time interaction was not significantly different. The 10% DBSFL hens had significantly higher percent monocytes (5.92 ± 0.18%) than the control hens (3.87 ± 0.26%) ($F(2,2.21) = 20.47$, $p = 0.04$). The percent lymphocytes of 18% DBSFL hens were 5.08 ± 0.18%.

Table 7. There was a significant difference in WBC differential counts between 20-week-old hens' baseline plasma and 35-week-old hens' blood samples (36 samples within 12 experimental units).

WBC Differential	Age		SEM	F Ratio	*p*-Value
	20-Week-Old	35-Week-Old			
WBC count ($\times 10^9$/L)	26.19	16.18	1.65	10.43	0.046
Hematocrit or packed cell volume (L/L)	0.30	0.26	0.006	56.97	0.004
Basophils (%)	1.24	2.57	0.53	26.97	0.02
Heterophils (%)	27.14	42.96	2.04	54.19	0.005
Lymphocytes (%)	63.17	50.58	1.26	25.54	0.01
Lymphocytes absolute ($\times 10^9$/L)	16.68	7.89	0.77	41.00	0.006
Heterophil:lymphocyte ratios (H:L ratio)	0.54	0.96	0.05	23.74	0.02
Monocytes (%)	7.36	2.55	0.50	24.33	0.01
Monocytes absolute ($\times 10^9$/L)	1.91	0.45	0.13	38.75	0.007

3.8. Fecal Microbial Analysis

Results of the fecal culture indicated no Salmonella, Shigella, or Campylobacter contamination in the trailers (or consequently, the treatments). Fecal culture direct exam results are presented in Table 8.

Table 8. Fecal microbial analysis direct exam results (n = 6).

Diets	Replication (Trailer)	Direct Exam [1] Gram-Negative Bacilli	Gram Positive Bacilli	Gram Positive Cocci
0% DBSFL [2]	1	Moderate	Moderate	Many
	2	Moderate	Scant	Moderate
10% DBSFL	1	Moderate	Moderate	Many
	2	Moderate	Scant	Moderate
18% DBSFL	1	Moderate	-	Moderate
	2	Moderate	Moderate	Many

[1] 'Many' indicates that the organism was present in all areas of the culture plates; 'moderate' indicates half the plate was covered; and 'scant' indicates that the organism was only present in the initial sample inoculum.
[2] DBSFL: Dried black soldier fly larvae.

3.9. Health and Welfare Parameters

There was neither mortality nor any observable signs of morbidity. Plumage conditions were excellent, and there were no wounds (scores 4 out of 4 with no variation) on any of the hens from all diet treatments.

3.10. Post-Mortem Examination of Gastrointestinal Tract

The 18% DBSFL hens showed a higher percent (of the digestive tract) of duodenum weight than control birds (Table 9).

Table 9. Post-mortem examinations of the gastrointestinal tract in hens on dried black soldier fly larvae (DBSFL) diets.

DI Weights [1]	Diet 0% DBSFL		10% DBSFL		18% DBSFL	
	Mean	SEM	Mean	SEM	Mean	SEM
Hen Live weight (kg)	1.93	0.08	1.73	0.04	1.73	0.10
Total DI Weight (g)	109.77	6.05	107.03	2.88	110.11	6.39
DW % BW [2]	5.73	0.33	6.19	0.11	6.43	0.43
Proventriculus % BW [2]	0.34	0.02	0.34	0.01	0.38	0.02
Proventriculus % DW [3]	5.87	0.12	5.52	0.12	5.96	0.08
Gizzard % BW	1.50	0.08	1.66	0.08	1.68	0.15
Gizzard % DW	26.27	0.69	26.90	1.15	26.03	0.90
Duodenum % BW	0.49	0.03	0.55	0.03	0.59	0.03
Duodenum % DW	8.48b	0.22	8.93ab	0.47	9.28a	0.20
Jejunum % BW	0.55	0.06	0.69	0.04	0.69	0.08
Jejunum % DW	9.45	0.57	11.19	0.59	10.59	0.66
Ileum % BW	0.38	0.03	0.49	0.02	0.48	0.06
Ileum % DW	6.65	0.28	7.98	0.30	7.39	0.51
Ceca % BW	0.39	0.03	0.38	0.01	0.37	0.03
Ceca % DW	6.81	0.29	6.19	0.32	5.80	0.20
Colon % BW	0.21	0.02	0.20	0.02	0.17	0.01
Colon % DW	3.69	0.27	3.24	0.30	2.66	0.18
Liver % BW	1.69	0.09	1.69	0.06	1.88	0.10
Liver % DW	29.65	0.38	27.26	0.77	29.47	1.01
Pancreas % BW	0.18	0.01	0.17	0.01	0.18	0.01
Pancreas % DW	3.16	0.14	2.80	0.16	2.83	0.14

[1] DI = Digestive system segments examined; [2] BW = hen live weight; [3] DW = DI weight.

4. Discussion

Changes in the production efficiency indicators of laying hens are expected when insects are included in their diets as substitute protein and lipid sources [9,10]. In the current study, control hens had a better weight gain, laid more and bigger eggs while consuming less feed, and had a better feed conversion ratio than 18% DBSFL hens. However, the production performance of the 10% DBSFL hens was not significantly different from the control hens in many of the parameters assessed (e.g., HDEP). Previous studies also recommended that the partial substitution of conventional protein sources in poultry diets with insects be feasible [12].

The effects of the use of BSFL meal in poultry feed have been investigated in several studies, while the use of full-fat BSFL has been reported only in a few studies. The fat content of the BSFL meal is extracted using different methods. For example, the larvae are dehydrated and ground, and then the Soxhlet method is used to extract the fat. Alternatively, the grounded larvae are repeatedly washed with a hexane organic solvent [37]. Another method is to use compression [18]. In addition, the inclusion rates and methods (i.e., whether they were used to substitute nutrients in the diet or used as an extra nutrient source alongside a formulated diet), BSFL production and processing methods, poultry production system, and hen age also varied among different studies, which all contribute to the reporting of inconsistent results in different articles. For example, Ruhnke et al. [38] reported that offering full-fat DBSFL ad libitum in an on-range choice feeding trial (from 47 to 62 weeks of age) did not affect performance parameters, but the egg weight was significantly lower in DBSFL hens. Nevertheless, Kawasaki et al. [39] found that the inclusion of full-fat BSFL and black soldier fly pre-pupae (from 24 to 5 weeks of age) did not impact the feed intake or hen egg-laying rates, while the pre-pupae treatment level resulted in the production of heavier eggs than the control and BSFL treatment levels. Moreover, Tahamtani et al. [40] fed four different amounts of live BSFL (0% to ad libitum) to laying hens (from 18 to 30 weeks of age) in addition to their full feed concentrate diet and found that hens (at ad libitum feeding) had higher protein, lipid, and energy intakes and weight gains compared to the other groups; nevertheless, they did not notice any effects on egg production or egg weight.

Similar inconsistencies have also been observed in the results reported from the BSFL meal inclusions in laying hen diets. Maurer et al. [41] used partly BSFL meal to replace soybean meal partially and fully in organic layer diets and fed old, spent White Leghorn layers (64–74 weeks old) for 3 weeks. They concluded that feed efficiency was maintained at a level similar to that of soybean meal-based diets, and the metabolic and health status of the hens were not affected. Marono et al. [42] studied the effects of substituting soybean meal completely with BSFL meal in the diet of 24 to 45-week-old hens. They found that feed intake, lay percentage, egg weight, and FCR declined in the BSFL group compared to the soybean meal-fed group.

The within-treatment live weight of 34-week-old hens in the control and the two DBSFL treatment groups in the current study all exceeded the target weight of 1.83 kg recommended by Novogen [31]. The 18% DBSFL hens consumed more feed compared to the other two groups, but the control hens gained more weight at a faster rate than the 10% and 18% DBSFL hens. As a result, control hens had significantly better FCR compared to the 18% DBSFL hens. Several dietary factors can potentially affect the FCR and performance of the 18% DBSFL hens, as listed below.

1. The 18% DBSFL diet was 30% bulkier than the other two diets because of the addition of 10% wheat bran to balance the nutrient provision with the high lipid content of DBSFL in the diet. The bulkiness of a diet impacts the amount of feed consumed (see Waldroup et al. [43]). Leeson et al. [44] reported increased feed intake because of nutrient dilution in bulky feed, which resulted in lower weight gain in layers. Oku et al. [45] demonstrated the inhibitory role of unavailable carbohydrates on the intestinal Ca absorption caused by the loss of Ca-binding protein when large quantities of undigested ingredients transit the gastrointestinal tract. Pelleting bulky diets reduces their bulkiness [46]. This solution can be applied to reduce the volume of full-fat DBSFL diets when using a bulkier ingredient (e.g., wheat bran) is necessary to balance the diets.
2. The exoskeleton of the chopped larvae and the tissue sheltered inside may pass through the digestive system without being digested, resulting in the loss of some

nutrients. Despines and Axtell [47] reported observations of undigested larval exoskeletons in the excreta of turkey pullets when they were fed darkling beetle larvae, and Zuidhof et al. [48] had similar observations in feeding house fly larvae to turkey pullets. The form of DBSFL (whole, chopped, or ground) therefore affects digestion and absorption of its nutrients [23,24]. Although crude protein levels were similar in the control and 18% BSFL diets, the plasma protein concentration was lower in 18% DBSFL hens, probably because of the lower digestibility and absorption of the nutrients in DBSFL considering the presence of the exoskeleton. Plasma proteins are good indicators of health and contribute to gluconeogenesis, maintenance of colloid osmotic pressure, production of enzymes and immunoglobulins, and transport of minerals and hormones [49]. Kerstetter et al. [50,51] reported the negative impact of low-dietary protein on Ca absorption in the gut. Plasma protein and Ca concentrations declined with increases in DBSFL levels in the present study. The 18% BSFL hens also had thinner egg shells compared to the control eggs [52]. Short-term studies and studies using old or low egg production hens may not be able to detect such Ca deficiency, and this may explain the inconsistency in the results reported from different studies.

3. Another potential factor affecting the Ca digestion and absorption of the 18% DBSFL hens could be related to the use of full-fat DBSFL. Using full-fat DBSFL to fully replace soybean meal and the majority of the soybean oil in the diet resulted in a high lipid content in the feed. The negative impact of the higher fatty acid concentrations in feed on the absorption of some nutrients, especially Ca, is well documented [53,54] because Ca can combine with fatty acids in the gastrointestinal tract to form insoluble soaps, which prevent its absorption. The availability of a high level of digestible Ca is crucial for egg production. Although the Ca concentrations were higher than the recommended level in all three experimental diets, the Ca present in the exoskeleton of the whole or chopped DBSFL could have a low bioavailability (see point 4 below). This may result in a faster depletion of the medullary bone Ca a few weeks after the beginning of egg production [55]. The drop in egg production of the 18% DBSFL hens contributed to the FCR calculation. In addition to the drop in egg production, the egg weights of the 18% DBSFL hens were also significantly lighter than the control eggs after the first 4 weeks of the experiment. Further research to determine the digestibility of the Ca in the DBSFL exoskeleton for laying hens may be worthwhile.
4. The total percent oil concentrations (crude fat) of the three experimental diets exceeded the fat requirements [30]. However, the plasma triglycerides of the 18% DBSFL hens were lower than those of the control hens, and Marono et al. [42] reported similar results. Avian energy reserves are mostly stored as lipids, specifically triglycerides [56]. Serum triglyceride concentrations are the most sensitive blood metabolites to feed or water deprivation [57]. Hossain and Blair [58] also reported that serum triglyceride concentrations were significantly lower in the birds fed diets containing chitin compared to the control birds. Lower plasma triglycerides in the 18% DBSFL hens may indicate that they were not absorbing the optimally required nutrients.
5. Finke [23] reported that the chitin content of the BSFL is 2-3% on an "as-is" basis (i.e., 5 to 7.5% on a dry-matter basis). Marono et al. [42] calculated that full replacement of soybean meal with 17% of BSFL meal (with 5.4% chitin content on a dry-matter basis) resulted in the provision of 1.02 g chitin/day/hen. The presence of chitin in the exoskeleton of DBSFL may influence the digestibility and bioavailability of its nutrients [59,60]. Chitin is insoluble in most solvents and has limited digestibility in poultry diets, and the inclusion of chitin in broiler feed significantly reduced the apparent digestibility of protein and the serum triglyceride concentration [58]. Chitin may also result in a slight overestimation of the crude protein content of the feed because it contains an N-acetyl group in its structure [61]. Biasato et al. [62] documented higher feed intake in diets containing high levels of chitin.

6. The results of the present study showed that the 18% DBSFL hens had a heavier duodenum (percent of the digestive tract) than the control hens. Johnson [55] indicated that Ca absorption formation occurs in the duodenum and upper jejunum. Biasato et al. [62] reported that inclusion of the yellow mealworm larvae in broiler chicken diets resulted in shorter villi and deeper crypts in the duodenum. This, in turn, decreased the digestibility and absorption of nutrients and caused poor performance in the birds. The full replacement of the soybean meal with BSFL meal resulted in a higher villi height in the duodenum but a lower villi height in the jejunum and the ileum in laying hens, as well as a change in the enzymatic activities of the brush border membrane [42,63]. The effects of the inclusion of BSFL on the digestibility of nutrients and intestinal enzymatic activities require further investigation.

Furthermore, one of the concerns about the use of DBSFL in poultry feed was related to the possibility of the presence of pathogens in feeds containing DBSFL. The inclusion of DBSFL in the present study did not introduce new pathogens, and the fecal microbial composition was comparable among the treatment levels. Another concern regarding the use of DBSFL has been the potential for higher concentrations of heavy metals in the DBSFL. The 18% DBSFL diet had the lowest concentrations of Cu and Cr compared to the other two diets. In addition, Zn concentrations were similar in all three diets. The As, Cd, and Pb concentrations of all three diets were negligible. The results of the present study revealed that the heavy metal contents of DBSFL were not significant, and none of the heavy metals examined exceeded the reported limits [64]. The heavy metal content of the DBSFL is affected by the sources of feed material provided to the larvae during their production period.

There was no mortality or any observable signs of health disorders in the flocks. Assessing plumage condition and wound scores, all the hens had top scores with no variation. An examination of the WBC counts did not reveal any signs of stress in the hens.

In the current study, hens from all groups had access to the cover crop in the free-range production system. The amount of cover crop consumption was not considered in the calculation of the FCR; however, it was controlled for by providing the same level of access to all groups at all times.

5. Conclusions

In summary, the use of DBSFL in poultry diets had no adverse effects on the health and welfare of the hens. Although 18% DBSFL hens were slower in gaining weight, gained less weight, and laid fewer and smaller eggs than the control hens, they were still gaining weight, maintaining above 90% HDEP, and their eggs were still in the Canadian large egg category.

The use of chopped full-fat DBSFL in the diet increased the bulkiness of the feed. Using BSFL meals may alleviate this problem. The BSFL meal will not be high in lipid content, so it will not be necessary to increase the bulkiness of the feed to balance the diet when substituting for soybean meal in the diets.

In conclusion, the authors recommend partial replacement of soybean meal and oil with full-fat DBSFL in layer diets because, at the partial substitution treatment level of 10% DBSFL, hen health and welfare, feed safety, and production efficiency parameters were comparable to those of the control hens. Further studies are required to determine the most efficient partial substitution rates for the inclusion of DBSFL in poultry diets and to develop new production efficiency criteria to facilitate the use of edible insects and alternative feed ingredients in poultry diets.

Supplementary Materials: The following is available online at https://www.mdpi.com/article/10.3390/agriculture14010031/s1, Figure S1: A photo of the trailer designed and built for the purposes of this study.

Author Contributions: Conceptualization, M.B. and K.M.C.; methodology, M.B. and K.M.C.; formal analysis, M.B.; investigation, M.B.; resources, M.B. and K.M.C.; data curation, M.B.; writing—original draft preparation, M.B.; writing—review and editing, M.B. and K.M.C.; visualization, M.B.; supervision, M.B. and K.M.C.; project administration, M.B. and K.M.C.; funding acquisition, M.B. and K.M.C. All authors have read and agreed to the published version of this manuscript.

Funding: The research was supported by funds from Egg Farmers of Canada (Ottawa, Ontario), Enterra Feed Corporation (Langley, British Columbia), the University of British Columbia Specialty Birds Research Committee, and a Mitacs Accelerate Scholarship.

Institutional Review Board Statement: The animal study protocol was approved by the University of British Columbia (UBC) Animal Care Committee (Certificate #A16-0227).

Data Availability Statement: The data presented in this study are available on request from the corresponding author.

Acknowledgments: Research facility support was provided by the UBC Avian Research Centre, the Kwantlen Polytechnic University Teaching and Education Centre (Richmond, BC), and the UBC Centre for Comparative Medicine (Laura Mowbray for providing veterinary assistance). We would like to thank staff and students of the Sustainable Agriculture and Food Systems Program at KPU, especially faculty members Rebecca Harbut and Mike Bomford and previous staff and students Torin Boyle, Kathy Whittemore, Eric Wirsching, Grace Augustinowicz, and Connor Sudbury for technical assistance. We thank Andrew Vickerson (Enterra Feed Corporation) for providing feedback on the proposal and research reports, and Rod Reid (In-Season Farms Inc., Abbotsford, BC) for preparing the experimental diets, supplying the experimental layers, and providing constructive inputs for raising free-range chickens. Robert Blair (UBC Applied Animal Biology), Jennifer Arthur (Agriculture and Agri-Food Canada), Frederick Silversides (formerly Agriculture and Agri-Food Canada), Darin Bennett (California Polytechnic State University), and the late Stewart Paulson (formerly British Columbia Ministry of Agriculture) critically reviewed earlier drafts of this manuscript.

Conflicts of Interest: The authors declare no conflicts of interest.

References

1. Falcon, W.P.; Naylor, R.L.; Shankar, N.D. Rethinking global food demand for 2050. *Popul. Dev. Rev.* **2022**, *48*, 921–957. [CrossRef]
2. Bikker, P.; Jansman, A.J.M. Composition and utilization of feed by monogastric animals in the context of circular food production systems. *Animal* **2023**, *17*, 100892. [CrossRef] [PubMed]
3. Ravindran, V.; Blair, R. Feed resources for poultry production in Asia and the Pacific. III. Animal protein sources. *World's Poult. Sci. J.* **1993**, *49*, 219–235. [CrossRef]
4. Jędrejek, D.; Levic, J.; Wallace, J.; Oleszek, W. Animal by-products for feed: Characteristics, European regulatory framework, and potential impacts on human and animal health and the environment. *J. Animal Feed Sci.* **2016**, *25*, 189–202. Available online: http://oa.fins.uns.ac.rs/handle/123456789/56 (accessed on 7 August 2023). [CrossRef]
5. Sapkota, A.R.; Lefferts, L.Y.; McKenzie, S.; Walker, P. What do we feed to food-production animals? A review of animal feed ingredients and their potential impacts on human health. *Environ. Health Perspect.* **2007**, *115*, 663. [CrossRef] [PubMed]
6. National Standard of Canada. Canadian General Standards Board guidelines for Organic production systems (CAN/CGSB-32.310-2015). *Amended 2021*. Available online: https://www.tpsgc-pwgsc.gc.ca/ongc-cgsb/programme-program/normes-standards/internet/032-310/032-310-eng.html (accessed on 10 August 2023).
7. Früh, B.; Schlatter, B.; Isensee, A.; Maurer, V.; Willer, H. Report on organic protein availability and demand in Europe. Research Institute of Organic Agriculture: Frick, Switzerland, 2014. Available online: https://orgprints.org/id/eprint/28067/3/FINAL-REPORT-ICOPP-2015-02-08.pdf (accessed on 13 May 2023).
8. FAO. Tackling food loss and waste: A triple win opportunity. 2022. Available online: https://www.fao.org/newsroom/detail/FAO-UNEP-agriculture-environment-food-loss-waste-day-2022/en (accessed on 5 September 2023).
9. Sogari, G.; Oddon, S.B.; Gasco, L.; van Huis, A.; Spranghers, T.; Mancini, S. Recent advances in insect-based feeds: From animal farming to the acceptance of consumers and stakeholders. *Animal* **2023**, *17*, 100904. [CrossRef] [PubMed]
10. Sverguzova, S.V.; Shaikhiev, I.H.; Sapronova, Z.A.; Fomina, E.V.; Makridina, Y.L. Use of fly larvae *Hermetia illucens* in poultry feeding: A review paper. *J. Water Land Dev.* **2021**, *49*, 95–103. [CrossRef]
11. Alagappan, S.; Rowland, D.; Barwell, R.; Mantilla, S.M.O.; Mikkelsen, D.; James, P.; Yarger, O.; Hoffman, L.C. Legislative landscape of black soldier fly (*Hermetia illucens*) as feed. *J. Insects Food Feed* **2022**, *8*, 343–355. [CrossRef]
12. Moula, N.; Detilleux, J. A meta-analysis of the effects of insects in feed on poultry growth performances. *Animals* **2019**, *9*, 201. [CrossRef]
13. Diener, S.; Zurbrügg, C.; Tocknera, K. Conversion of organic material by black soldier fly larvae–Establishing optimal feeding rates. *Waste Manag. Res.* **2009**, *27*, 603–610. [CrossRef]
14. Tran, G.; Gnaedinger, C.; Mélin, C. Black soldier fly larvae (*Hermetia illucens*). Feedipedia, a programme by INRA, CIRAD, AFZ and FAO. 2015. Available online: http://www.feedipedia.org/node/16388 (accessed on 12 August 2023).

15. Belghit, I.; Liland, N.S.; Gjesdal, P.; Biancarosa, I.; Menchetti, E.; Li, Y.; Waagbø, R.; Krogdahl, Å.; Lock, E.-J. Black soldier fly larvae meal can replace fish meal in diets of sea-water phase Atlantic salmon (*Salmo salar*). *Aquaculture* **2019**, *503*, 609–619. [CrossRef]
16. Borgogno, M.; Dinnella, C.; Iaconisi, V.; Fusi, R.; Scarpaleggia, C.; Schiavone, A.; Monteleone, E.; Gasco, L.; Parisi, G. Inclusion of *Hermetia illucens* larvae meal on rainbow trout (*Oncorhynchus mykiss*) feed: Effect on sensory profile according to static and dynamic evaluations. *J. Sci. Food Agric.* **2017**, *97*, 3402–3411. [CrossRef] [PubMed]
17. Driemeyer, H. Evaluation of black soldier fly (*Hermetia illucens*) larvae as an alternative protein source in pig creep diets in relation to production, blood and manure microbiology parameters. Ph.D. Thesis, Stellenbosch University, Stellenbosch, South Africa, 2016.
18. Kroeckel, S.; Harjes, A.G.; Roth, I.; Katz, H.; Wuertz, S.; Susenbeth, A.; Schulz, C. When a turbot catches a fly: Evaluation of a pre-pupae meal of the Black Soldier Fly (*Hermetia illucens*) as fish meal substitute—Growth performance and chitin degradation in juvenile turbot (Psetta maxima). *Aquaculture* **2012**, *364*, 345–352. [CrossRef]
19. Lopez, Z.P.; Purnamasari, L.; dela Cruz, J.F. A review: Evaluation of black soldier fly (*Hermetia illucens*) larvae meal as a dietary protein source in poultry diets. *Biotropika J. Trop. Biol.* **2022**, *10*, 191–202. [CrossRef]
20. Sealey, W.M.; Gaylord, T.G.; Barrows, F.T.; Tomberlin, J.K.; McGuire, M.A.; Ross, C.; St-Hilaire, S. Sensory analysis of rainbow trout, Oncorhynchus mykiss, fed enriched black soldier fly prepupae, *Hermetia illucens. J World Aquac Soc.* **2011**, *42*, 34–45. [CrossRef]
21. St-Hilaire, S.; Cranfill, K.; McGuire, M.A.; Mosley, E.E.; Tomberlin, J.K.; Newton, L.; Sealey, W.; Sheppard, C.; Irving, S. Fish offal recycling by the black soldier fly produces a foodstuff high in omega-3 fatty acids. *J. World Aquac. Soc.* **2007**, *38*, 309–313. [CrossRef]
22. Stadtlander, T.; Stamer, A.; Buser, A.; Wohlfahrt, J.; Leiber, F.; Sandrock, C. *Hermetia illucens* meal as fish meal replacement for rainbow trout on farm. *J. Insects Food Feed.* **2017**, *3*, 165–175. [CrossRef]
23. Finke, M.D. Complete nutrient content of four species of feeder insects. *Zoo Biol.* **2013**, *32*, 27–36. [CrossRef]
24. Blair, R. *Nutrition and Feeding of Organic Poultry*; CABI: Wallingford, UK, 2008.
25. Erickson, M.C.; Islam, M.; Sheppard, C.; Liao, J.; Doyle, M.P. Reduction of *Escherichia coli O157: H7* and *Salmonella enterica serovar enteritidis* in chicken manure by larvae of the black soldier fly. *J. Food Prot.* **2004**, *67*, 685–690. [CrossRef]
26. Lalander, C.; Diener, S.; Magri, M.E.; Zurbrügg, C.; Lindström, A.; Vinnerås, B. Faecal sludge management with the larvae of the black soldier fly (*Hermetia illucens*)—From a hygiene aspect. *Sci. Total Environ.* **2013**, *458*, 312–318. [CrossRef]
27. Carpenter, K.J.; Clegg, K.M. The metabolizable energy of poultry feeding stuffs in relation to their chemical composition. *J. Sci. Food Agric.* **1956**, *7*, 45–51. [CrossRef]
28. NRC. *Nutrient Requirements of Poultry*, 9th ed.; The National Academies Press: Washington, DC, USA, 1994.
29. AOAC. *Official Methods of Analysis of AOAC International*; Latimer, G.W., Ed.; AOAC International: Gaithersburg, MD, USA, 2012.
30. Novogen. *Management Guide Commercial Layers—NovoGen Brown*; NovoGen: Quintin, France, 2015.
31. Novogen. *Production Targets Commercial Layers—Alternative Systems—NOVOgen Brown*; NovoGen: Quintin, France, 2015.
32. Tauson, R.; Kjaer, J.; Maria, G.A.; Cepero, R.; Holm, K.E. Applied scoring of integument and health in laying hens. *Animal Sci. Pap. Rep.* **2005**, *23* (Suppl. S1), 153–159.
33. Turk, D.E. The anatomy of the avian digestive tract as related to feed utilization. *Poult. Sci.* **1982**, *61*, 1225–1244. [CrossRef] [PubMed]
34. Schank, J.C.; Koehnle, T.J. Pseudoreplication is a pseudoproblem. *J. Comp. Psychol.* **2009**, *123*, 421–433. [CrossRef] [PubMed]
35. Olejnik, S.; Algina, J. Generalized eta and omega squared statistics: Measures of effect size for some common research designs. *Psychol. Methods* **2003**, *8*, 434–447. [CrossRef] [PubMed]
36. Cohen, J. A power primer. *Psychol. Bull.* **1992**, *112*, 155–159. [CrossRef] [PubMed]
37. Monteiro dos Santos, D.K.; Santana, T.M.; de Matos Dantas, F.; Farias, A.B.D.S.; Epifânio, C.M.F.; Prestes, A.G.; da Fonseca, F.A.L.; Parisi, G.; Viegas, E.M.M.; Gonçalves, L.U. Defatted black soldier fly larvae meal as a dietary ingredient for tambaqui (*Colossoma macropomum*): Digestibility, growth performance, haematological parameters, and carcass composition. *Aquac. Res.* **2022**, *53*, 6762–6770. [CrossRef]
38. Ruhnke, I.; Normant, C.; Campbell, D.L.; Iqbal, Z.; Lee, C.; Hinch, G.N.; Roberts, J. Impact of on-range choice feeding with black soldier fly larvae (*Hermetia illucens*) on flock performance, egg quality, and range use of free-range laying hens. *Anim Nutr.* **2018**, *4*, 452–460. [CrossRef]
39. Kawasaki, K.; Hashimoto, Y.; Hori, A.; Kawasaki, T.; Hirayasu, H.; Iwase, S.I.; Hashizume, A.; Ido, A.; Miura, C.; Miura, T.; et al. Evaluation of black soldier fly (*Hermetia illucens*) larvae and pre-pupae raised on household organic waste, as potential ingredients for poultry feed. *Animals* **2019**, *9*, 98. [CrossRef]
40. Tahamtani, F.M.; Ivarsson, E.; Wiklicky, V.; Lalander, C.; Wall, H.; Rodenburg, T.B.; Tuyttens, F.A.M.; Hernandez, C.E. Feeding live Black Soldier Fly larvae (*Hermetia illucens*) to laying hens: Effects on feed consumption, hen health, hen behavior, and egg quality. *Poult. Sci.* **2021**, *100*, 101400. [CrossRef]
41. Maurer, V.; Holinger, M.; Amsler, Z.; Früh, B.; Wohlfahrt, J.; Stamer, A.; Leiber, F. Replacement of soybean cake by *Hermetia illucens* meal in diets for layers. *J. Insects Food Feed.* **2015**, *2*, 83–90. [CrossRef]
42. Marono, S.; Loponte, R.; Lombardi, P.; Vassalotti, G.; Pero, M.E.; Russo, F.; Gasco, L.; Parisi, G.; Piccolo, G.; Nizza, S.; et al. Productive performance and blood profiles of laying hens fed *Hermetia illucens* larvae meal as total replacement of soybean meal from 24 to 45 weeks of age. *Poult. Sci.* **2017**, *96*, 1783–1790. [CrossRef] [PubMed]

43. Waldroup, P.W.; Mitchell, R.J.; Payne, J.R.; Johnson, Z.B. Characterization of the response of broiler chickens to diets varying in nutrient density content. *Poult. Sci.* **1976**, *55*, 130–145. [CrossRef]
44. Leeson, S.; Summers, J.D.; Caston, L.J. Response of layers to low nutrient density diets. *J. Appl. Poult. Res.* **2001**, *10*, 46–52. [CrossRef]
45. Oku, T.; Konishi, F.; Hosoya, N. Mechanism of inhibitory effect of unavailable carbohydrate on intestinal calcium absorption. *J. Nutr.* **1982**, *112*, 410–415. [CrossRef] [PubMed]
46. Senkoylu, N.; Dale, N. Sunflower meal in poultry diets: A review. *World's Poult. Sci. J.* **1999**, *55*, 153–174. [CrossRef]
47. Despines, J.L.; Axtell, R.C. Feeding behaviour and growth of turkey poults fed larvae of darkling beetle *Alphitobius diaperinus*. *Poult. Sci.* **1994**, *73*, 1526–1533. [CrossRef]
48. Zuidhof, M.J.; Molnar, C.L.; Morley, F.M.; Wray, T.L.; Robinson, F.E.; Khan, B.A.; Al-Ani, L.; Goonewardene, L.A. Nutritive value of house fly (*Musca domestica*) larvae as a feed supplement for turkey poults. *Anim. Feed Sci. Tech.* **2003**, *105*, 225–230. [CrossRef]
49. Crespo, R.; Shivaprasad, H.L. Interpretation of Laboratory Results and Values. In *Backyard Poultry Medicine and Surgery: A Guide for Veterinary Practitioners*; Greenacre, C.B., Morishita, T.Y., Eds.; John Wiley and Sons: Ames, IA, USA, 2014; pp. 283–296.
50. Kerstetter, J.E.; O'Brian, K.O.; Insogna, K.L. Dietary protein affects intestinal calcium absorption. *Am. J. Clin. Nutr.* **1998**, *68*, 859–865. [CrossRef]
51. Kerstetter, J.E.; O'Brien, K.O.; Insogna, K.L. Dietary protein, calcium metabolism, and skeletal homeostasis revisited. *Am. J. Clin. Nutr.* **2003**, *78*, 584S–592S. [CrossRef]
52. Bejaei, M.; Cheng, K.M. The effect of including full-fat dried black soldier fly larvae in laying hen diet on egg quality and sensory characteristics. *J. Insects Food Feed.* **2020**, *6*, 305–314. [CrossRef]
53. Gacs, G.; Barltrop, D. Significance of Ca-soap formation for calcium absorption in the rat. *Gut* **1977**, *18*, 64–68. [CrossRef] [PubMed]
54. Bouvarel, I.; Nys, Y.; Lescoat, P. Hen nutrition for sustained egg quality. In *Improving the Safety and Quality of Eggs and Egg Products*; Nys, Y., Bain, M., Van Immerseel, F., Knovel, A., Eds.; Woodhead Pub: Oxford, UK, 2011; pp. 261–299.
55. Johnson, A.L. Reproduction in the female. In *Sturkie's Avian Physiology*; Whittow, G.C., Ed.; Academic Press: Cambridge, MA, USA, 1999; pp. 569–596.
56. Blem, C.R. Energy Balance. In *Sturkie's Avian Physiology*; Whittow, G.C., Ed.; Academic Press: Cambridge, MA, USA, 1999; pp. 327–342.
57. Huff, W.E.; Bayyari, G.R.; Roth, N.C.; Balog, J.M. Effect of feed and water withdrawal on green liver discoloration, serum triglycerides, and hemoconcentration in turkeys. *Poult. Sci.* **1996**, *75*, 59–61. [CrossRef] [PubMed]
58. Hossain, S.M.; Blair, R. Chitin utilisation by broilers and its effect on body composition and blood metabolites. *Br. Poult. Sci.* **2007**, *48*, 33–38. [CrossRef] [PubMed]
59. Razdan, A.; Pettersson, D. Effect of chitin and chitosan on nutrient digestibility and plasma lipid concentrations in broiler chickens. *Br. J. Nutr.* **1994**, *72*, 277–288. [CrossRef] [PubMed]
60. Marono, S.; Piccolo, G.; Loponte, R.; Di Meo, C.; Attia, Y.A.; Nizza, A.; Bovera, F. In vitro crude protein digestibility of *Tenebrio molitor* and *Hermetia illucens* insect meals and its correlation with chemical composition traits. *Ital. J. Anim. Sci.* **2015**, *14*, 338–343. [CrossRef]
61. Finke, M.D. Nutrient content of insects. In *Encyclopedia of Entomology*; Springer: Dordrecht, The Netherlands, 2008; pp. 2623–2646.
62. Biasato, I.; Gasco, L.; De Marco, M.; Renna, M.; Rotolo, L.; Dabbou, S.; Capucchio, M.T.; Biasibetti, E.; Tarantola, M.; Sterpone, L.; et al. Yellow mealworm larvae (*Tenebrio molitor*) inclusion in diets for male broiler chickens: Effects on growth performance, gut morphology, and histological findings. *Poult. Sci.* **2018**, *97*, 540–548. [CrossRef]
63. Cutrignelli, M.I.; Messina, M.; Tulli, F.; Randazzo, B.; Olivotto, I.; Gasco, L.; Loponte, R.; Bovera, F. Evaluation of an insect meal of the Black Soldier Fly (*Hermetia illucens*) as soybean substitute: Intestinal morphometry, enzymatic and microbial activity in laying hens. *Res. Vet. Sci.* **2018**, *117*, 209–215. [CrossRef]
64. Zhang, F.; Li, Y.; Yang, M.; Li, W. Content of heavy metals in animal feeds and manures from farms of different scales in Northeast China. *Int. J. Environ. Res. Public Health* **2012**, *9*, 2658–2668. [CrossRef]

agriculture

Systematic Review

Cultivation of Black Soldier Fly (*Hermetia illucens*) Larvae for the Valorization of Spent Coffee Ground: A Systematic Review and Bibliometric Study

Donald John Calvien Hutabarat [1,2] and Dave Mangindaan [1,3,*]

1 Waste-Food-Environmental Nexus Research Interest Group, Bina Nusantara University, Jl. KH Syahdan No. 9, Jakarta 11480, Indonesia; donald.calvien@binus.ac.id
2 Food Technology Department, Faculty of Engineering, Bina Nusantara University, Jl. KH Syahdan No. 9, Jakarta 11480, Indonesia
3 Civil Engineering Department, Faculty of Engineering, Bina Nusantara University, Jl. KH Syahdan No. 9, Jakarta 11480, Indonesia
* Correspondence: dave.mangindaan@binus.ac.id

Abstract: The global surge in coffee consumption has led to the generation of significant amounts of spent coffee grounds (SCG), a by-product of the brewing process. If it is left unprocessed in the landfill, it will generate methane, one of the greenhouse gases, and therefore accelerate global warming. The intersection of SCG and its potential as a substrate for black soldier fly (BSF) larvae cultivation as one of the pathways for processing SCG becomes intriguing as we seek sustainable waste management solutions. The combination of both nutrition and toxic alkaloids (caffeine) makes SCG and/or other coffee parts intriguing for recycling (or upcycling) via BSF cultivation to generate insect protein. Due to its remarkable capacity to bioconvert organic waste into high-value proteins and fats, the black soldier fly, *Hermetia illucens*, has garnered attention in waste management and animal feed production. This comprehensive review sheds light on the recent development of using SCG as a substrate for BSF larvae.

Keywords: black soldier fly; larvae; spent coffee ground; coffee; coffee silverskin; bibliometric; systematic literature review

1. Introduction

1.1. Black Soldier Fly (BSF)

BSF larvae (*Hermetia illucens*) consume a wide range of organic waste. With a feed assimilation rate of 1.0–1.5 g g^{-1} day^{-1} [1–4], BSF larvae greatly assist in the reduction of the volume of waste, ensuring a more sustainable and effective waste management strategy. BSF larvae have been identified as efficient decomposers of organic waste. Several attributes and studies strengthen the superiority of BSF larvae in reducing organic waste such as fruit and vegetable waste, manure, meat waste, etc. into high-quality protein and lipids, which can be used as feedstock for poultry, fish, or swine, presenting an economic opportunity [5] and therefore making them versatile for waste management [6].

Owing to their rapid consumption of large amounts of organic waste, BSF larvae convert waste materials into their body mass, leaving behind significantly reduced waste volume [7,8]. More interestingly, they not only significantly reduce the amount of waste but are also able to reduce pathogen counts such as *Escherichia coli* and *Salmonella* sp., making the end product safer for disposal or use as soil amendment [9]. Thus, the utilization of BSF larvae in waste management is not just a method of waste reduction but a holistic approach to waste valorization, pathogen reduction, and environmental protection. Their ability to effectively reduce and valorize organic waste underscores the need for more widespread adoption and research into the BSF lifecycle and its integration into sustainable waste management systems.

BSF larvae possess high nutritional value and are rich in proteins as well as fats and other nutrients, making them an excellent source of feed for poultry, fish, and even pets. Their high nutritional content is rooted in their unique composition, which includes proteins, lipids, and other essential micronutrients, such as proteins (around 42–63%) [10], lipids (about 15–35%) [11], essential amino acids (often comparable to those of fish meal, a standard high-quality protein source) [12], minerals (such as calcium, phosphorus, magnesium, and potassium), and several vitamins [13].

Moreover, BSF larvae also have some antimicrobial peptides (that can be beneficial for the animals consuming them, potentially improving gut health and disease resistance [14]), as well as chitin (a polysaccharide that has potential benefits such as supporting the immune system and promoting gut health in animals) [15]. The nutritional profile of BSF larvae highlights their potential as an alternative and sustainable protein source for animal feed [5]. As the global demand for animal feed continues to increase, the high nutritional content of BSF larvae, coupled with their environmental benefits, makes them a compelling option for future feed formulations.

In addition to the aforementioned excellence of BSF larvae, by consuming and degrading organic waste as well as livestock manure, BSF larvae also simultaneously bioconvert greenhouse gases, such as methane and nitrous oxide, that might be produced from unprocessed and untreated organic wastes [16–18]. On the other hand, animal feed, especially fishmeal and soy, is associated with high GHG emissions during its production. BSF larvae can be used as an alternative protein source, reducing the demand for such feeds [19].

As BSF larvae consume organic carbon during their growth, a significant portion of this carbon is stored in their bodies, thus serving as a form of short-term carbon sequestration [17]. Furthermore, their frass can also be utilized as a soil amendment that serves as an alternative to synthetic fertilizers associated with high GHG emissions [20]. The high lipid content in BSF larvae can be extracted and converted into biodiesel. Producing biodiesel from BSF larvae can reduce dependence on fossil fuels, thereby offsetting the associated GHG emissions [5,21]. Incorporating BSF larvae into waste management and agricultural systems can serve as a multifaceted strategy to reduce GHG emissions. From diverting waste from landfills to providing sustainable feed alternatives, the benefits of BSF are evident. Adopting such innovative solutions will be vital in the ongoing quest to combat climate change and its impactful damages.

1.2. Spent Coffee Grounds (SCG) as a Potential Organic Substrate

SCG is the byproduct remaining after the brewing process of coffee beans. Over the past years, there has been growing interest in SCG due to its potential applications in various sectors, ranging from bioenergy production to food supplementation [22] and also in nanotechnology [23–26]. SCG is rich in dietary fibers (predominantly cellulose, hemicellulose, and lignin) [27]. It also has a reasonable amount of antioxidants (chlorogenic acid, of 10.7% to 23.82%, with the yield of extracts from coffee parts (pulp, husk, silverskin, spent waste) of 11.7% to 25.0%) [28] for potential use as nutraceuticals or food supplements, and also lipids (around 8–20% of SCGs dry weight [29], as a prospective candidate for biodiesel production [30,31]). However, it should be noted that SCG, as organic waste, if left without proper treatment (such as dumped in landfills), will generate CH_4 as one of the greenhouse gases and an agent of global warming [32,33].

Furthermore, SCG is also a source of minerals, namely potassium, phosphorus, magnesium, and calcium [34], and also some antioxidants and bioactive compounds such as chlorogenic acids (a major group of polyphenolic compounds in coffee, with antioxidant, anti-inflammatory, antiviral, and antihyperglycemic activities, 4–8% of dry weight) [35], melanoidins (about 25% of dry weight, responsible for the color of roasted coffee, possess antioxidant and antimicrobial properties) [36], and especially caffeine (around 1% of SCGs dry weight) [35,37]. Given the vast amounts of SCG produced worldwide due to the popularity of coffee, understanding its composition is crucial. This knowledge not only helps divert waste from landfills but also unlocks potential sustainable applications in various industries, especially for BSF-driven bioconversion.

1.3. The Effects of Feeding SCG to BSF

BSF has emerged as a significant player in the domain of waste management and for producing high-quality animal feed [38–40]. By modifying the nutritional inputs [41–44] of the BSF, researchers can control the quality and yield of its larvae, which serve as protein sources. One promising nutritional input is spent coffee grounds (SCG) [45–49]. We aim to review the influence of SCG in BSF feed, considering the effect on growth rate, nutritional profile, waste degradation, and the potential benefits and challenges. One challenge encountered was the caffeine content in SCG. High caffeine concentrations could pose toxicity risks to BSF larvae and, subsequently, to the animals consuming them. Thus, finding the optimal ratio of SCG in the feed mix is crucial for ensuring maximum yield and nutritional quality. To the best of our knowledge, this kind of study assessing the optimum composition of SCG in the feed for BSF is not available yet.

2. Materials and Methods

This study employs a systematic literature review and is supplemented with bibliometric analysis based on the analysis from the Scopus database (https://scopus.com, accessed on 1 October 2023). It is one of the largest databases utilized for reviews and bibliometric analysis due to its vast array of abstracts and citations from a multitude of journals, conferences, and books, thoroughly curated and inspected by a Content Selection and Advisory Board [50,51]. From the Scopus web, the keyword "illucens" was inputted. The usage of this single keyword in the scientific name of BSF is practiced because there is little difference in the results obtained from inputting "illucens" and "black soldier fly" [5].

We conducted a systematic review following the Preferred Reporting Items for Systematic Reviews and Meta-Analyses (PRISMA) statement. The PRISMA diagram for this study (accessed October 2023) is shown in Figure 1. Initially, there were 1683 titles shown under the "illucens" keyword. The results were filtered for the new recent titles from the last decase (2014–2023) to 1607 titles. These titles of BSF research were further sieved down to 22 titles with relation to SCG, or coffee, or coffee-related products. Moreover, those with minimum adherence were removed, and therefore there are 18 key papers to be further discussed in this study.

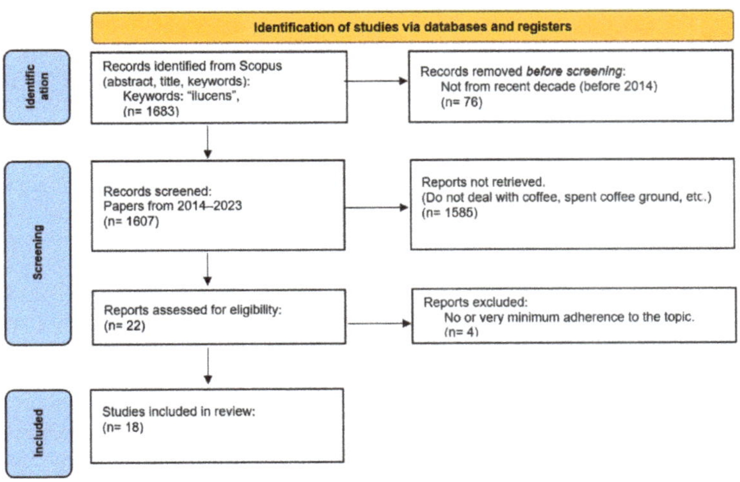

Figure 1. Preferred Reporting Items for Systematic Reviews and Meta-Analyses (PRISMA) for this study.

In addition to the systematic literature review about the effects of SCG on the growth of BSF, this paper also employed bibliometric mapping/analysis. The aforementioned 18 core papers, along with their essential keywords (provided by the authors), were analyzed using VOSViewer software (version 1.6.17) to provide visualization and mapping of the interconnection for this research. The mapping will be very useful to the development of future research directions, either for BSF cultivation, SCG utilization, or a combination of both.

3. Results

The state-of-the-art of the effects of SCG (and/or other parts of coffee) on BSF larvae cultivation is shown in Table 1. Furthermore, the effects of SCG-fed BSF larvae as a feed for fishes are shown in Table 2. Table 1 contains 12 entries, while Table 2 has 6 entries, totaling 18 reports as a result of the PRISMA procedure stated in Figure 1.

Table 1 displays various research results conducted to optimize the rearing of BSF larvae that focus on different rearing substrates, conditions, and their effects on larval growth and development. There are several research works in Table 1, such as feeding BSF (1) directly with coffee or spent coffee ground, (2) with algae-enriched coffee or coffee silverskin, (3) with various carbohydrate or protein sources (either mixed with coffee or not), and also (4) the application of SCG-derived BSF frass. There are some mixed results from Table 1, which indeed make this kind of research exciting.

3.1. *Cultivation of BSF Larvae Using SCG and/or Coffee Parts*

In Table 1, Permana and coworkers et al. [52] investigated the use of SCG with 7-day-old BSF larvae. Different feeding rates are tested, namely 12.5, 25, 50, 100, and 200 mg/larvae/day. Their study concluded that the optimal feeding rate is 200 mg/larvae/day, where it leads to a development time of around 25.6 days, efficient digestion conversion of 5%, pupae biomass of 14.8 mg, growth rate of 1.41 mg/day, waste reduction index of 0.997, and pupae protein content of up to 33.45%. In other research, Ospina-Granobles and Carrejo-Gironza used the Castillo variety of *Coffea arabica* to feed BSF larvae [53]. From their study, it is evident that different feeding rates (100, 160, and 200 mg/larvae/day) impacted the weight reduction index, larval development time, and the efficiency of the conversion of ingested food. With 100 mg/larvae/day, the reduction percentage of coffee pulp is 62.88% (wet basis), and the efficiency of conversion of ingested food = 7.89%. For the feeding of BSF using coffee at 160 mg/larvae/day, it achieves the highest weight of 115.9 mg with the shortest average development time of 38.65 days.

Truzzi et al. [54] detailed the use of coffee silverskin enriched with *Schizochytrium* sp. algae. The best results are observed with a 10% inclusion rate of *Schizochytrium* sp., enhancing the lipid and protein content in prepupae and increasing the levels of omega-3 fatty acids. Perhaps their study is the only report of the fatty acid profile of BSF larvae reared with coffee products. Based on their results, the majority of the fatty acids (with >10 g/100 g fatty acids) of BSF larvae fed with coffee silverskin enriched with algae generally are C12:0, C16:0, C18:0, and C18:1n9 (lauric acid, palmitic acid, stearic acid, and elaidic acid (trans-9-octadecanoic acid), respectively). However, for those enriched with *Schizochytrium* sp., C20:5n3 and C22:6n3 emerge (eicosapentaenoic acid and docosahexanoic acid, respectively). Those enriched with *Isochrysis* sp. do not show any new fatty acids with a composition of >10 g/100 g fatty acids. Similar to the fatty acid profile of BSF larvae fed with coffee silverskin, those reared with a total mix diet (composed of ground corn, barley, bran, alfalfa, wheat, etc.) [55] also show C12:0 and C16:0 fatty acids (lauric acid and palmitic acid, respectively) as the dominant ones (>10 g/100 g total defatted fatty acid (TDFA)). The BSF larvae fed a total mix diet also have C14:0 (myristic acid) as the majority of fatty acids, besides C12:0 and C16:0. On the other hand, it is also detected that there are C18:1 cis-9 and C18:2n-6 fatty acids (oleic acid and linoleic acid, respectively), although the concentration is quite moderate (3.69–8.5 g/100 g TDFA). Based on the comparison of profiles of fatty acids, it can be concluded that the profiles of fatty acids of BSF larvae are varied as a function of the substrates for the rearing process.

On the other hand, Milanovic et al. [56] also used the algae-enriched coffee as the feed for BSF. The algae are *Schizochytrium limacinum* and *Isochrysis galbana*. The results indicate a high prevalence of tetracycline resistance genes and an impact on antibiotic resistance (AR) gene distribution in larvae and frass. In another study about algae-enriched coffee silverskin, Osimani et al. [57] mapped different bacterial dominances in the larvae's microbiota and the frass. As identical algae are used (*Schizochytrium limacinum* and *Isochrysis galbana*), they present different microorganism communities. BSF reared with coffee silverskin with no algae enrichment showed the larvae's microbiota was dominated by *Paenibacillus*. bacteria are still present in the *S. limacinum*- and *I. galbana*-enriched coffee silverskin, but with the presence of other genera in the larvae's microbiota such as *Enterococcus, Lysinibacillus*, and *Morganella* (for *S. limacinum*-enriched silverskin) and even high abundances of *Brevundimonas, Enterococcus*, and *Paracoccus* (*I. galbana*-enriched silverskin). In the BSFs frass, *Brevundimonas* are detected for both BSF reared with algae-enriched silverskin, with the addition of *Alcaligenes* genera specifically for that of *I. galbana*-enriched coffee silverskin.

As the work by Osimani et al. [57] might be the only study of microorganism mapping of BSF reared with coffee products (coffee silverskin enriched with algae (*Schizochytrium limacinum, Isochrysis galbana*), it is highlighted that they have viable counts of lactic acid bacteria (*Lactococcus, Lactobacillus, Enterococcus, Weissella, Leuconostoc*) [57], as coffee byproducts are dominated by lactic acid bacteria according to a study by Pothakos and coworkers [58]. Furthermore, they also noted that the frass from insects is dominated by saprophytic microorganisms [57]. However, although the reports discussed in Table 1, entries 3–5, are from the same group that disclosed the details of the parameters of the BSF larvae cultivation, the weight evolution of the BSF larvae in their reports is not available. They just harvested or collected the larvae when the tegument color changed from white to black, without mentioning any specific duration.

Hadj Saadoun et al. tested several substrates (brewery-spent grains, tomato peels and seeds, cow's milk whey, grape stalk, bread dough, and SCG) for BSF prepupae [59]. Brewery-spent grains showed the best larval performance (0.22 g/prepupae). Other larval performances are 0.19 g/prepupae (tomato peels), and 0.14 g/prepupae (cheese). However, there is no development for SCG-fed BSF larvae (they died after 15 days, or survival rate = 0%, due to the high content of indigestible fibers and caffeine as a toxic alkaloid) [60]. Moreover, the highest ratio between unsaturated and saturated fatty acids is 1.34 (cow's milk and whey), while the lowest ratio is 0.37 (tomato peels and seeds).

Fischer et al. (Table 1, entry 7) experimented with SCG, donut dough, and a blend of SCG and dough (1:1) for BSF larvae and prepupae for 35 days, observing variations in survival, growth, and nutritional content [45]. Their best result is obtained from the feeding of BSF using SCG:donut dough 1:1, with a survival rate of 81%, with its frass containing the highest nitrogen (~4.20% dry matter (DM), compared to other feeding experiments. The result for feeding SCG:donut 1:1 is comparable to soybean meal and organic fertilizers. Feeding with sole SCG provided a survival rate of 45%, and surprisingly, that of the donut dough is much lower, with a survival rate of 24%. Although BSF larvae fed with donut dough are the longest (21.44 mm) and the largest (0.23 g), their net production is the lowest (0.60 ± 1.01 g/day/m^3). On the other hand, those fed with SCG:donut 1:1 have similar length (19.12 mm) and weight (0.18 g) but a net production of 4.42 ± 1.02 g/day/m^3. Based on the results, it can be concluded that the mixture of SCG and donut dough works in synergy by covering each other's drawbacks.

Sideris et al. compared SCG, brewer's spent grain (BSG), and a mix of both as substrates. BSG alone or in combination showed better results than SCG alone (larvae underperformed, with a very long development time = 35 days and a larval yield of <0.77 g in DM) [61]. The best result (for aspects of substrate mass reduction, protein conversion rate, and bioconversion rate) is acquired for feeding with reference feed (BΩ-321 feed,

Viozois S.A., Greece) that is 67.38% in substrate mass reduction, 45.77% protein conversion rate, and 23.28% bioconversion rate, besides the fast development time of 11 days only. Although the performance is lower, the usage of SCG, BSG, and brewer's yeast (BY) provides an alternative feed from waste by-products from food industries that will drive sustainability with comparable performance to that of the commercial feed (18.48–44.80% substrate mass reduction, 17.55–30.15 protein conversion rate, 6.25–17.22% bioconversion rate, and 8–17 days of development time).

Villa et al. (Table 1, entry 9) explored a mixture of sludge (S), brewery spent grains (BSG), coffee waste (C), and whey (W) with various compositions as a feed to BSF [46]. The mixtures affected larval development, size, and weight differently. With S:BSG:C:W = 50:10:10:30, BSF larvae are not developed (final height after 23 days = ~6 mm, final weight $\leq$ 10 mg). When the composition is tuned to S:BSG:C:W = 50:10:30:10 (increased portion of coffee waste), the larvae size and weight are significantly different (p-value < 0.05) when compared to the control (final height after 23 days = ~15–18 mm, final weight = 81.9 $\pm$ 5.3 mg). However, when coffee waste and waste are eliminated, leaving S:BSG = 50:50, or brewery spent grain is increased to 30%, with S:BSG:C:W = 50:30:10:10, then no significant difference is observed for larvae size and weight when compared to the control (final weight after 23 days is still between 15 and 18 mm, and final weight = ~90 mg).

Romano et al. examined the nutritional profile of BSF larvae reared for two weeks on sweet potato, SCG, or dough [47]. Each substrate affected the larve's nutritional content and bioconversion efficiency differently. For BSF larvae fed with sweet potato and with SCG (separately), the length and weight of the larvae and prepupae are in a similar range of around 17 mm and 17 g, respectively. It is shown in Table 1, entry 10, that those fed with SCG show significantly higher crude protein (~65% dry weight), with the highest composition of palmitic acid (24.85%) and linoleic acid (23.30%) compared to those fed with sweet potato (17.45%, 12.40%, respectively) or dough (15.21%, 18.92%, respectively). It also produces the highest total amino acid content (39.49 $\pm$ 2.57%). However, those fed with sweet potatoes have the strongest growth and nutritional profile of BSF larvae. This is due to the fact that its percentage of prepupae is similar to that of those fed with SCG (in the range of 36%) and higher than that of those fed with dough (prepupae percentage of 15%). On the other hand, those with sweet potatoes also have similar larval length, larval weight, and total amino content to those fed with SCG. However, as it has the highest lauric acid content (43%) that is considered to have anti-obesity properties for mammals, BSF larvae fed with sweet potato are considered to be the best compared to those fed with SCG (lauric acid content of only 27%).

As previously mentioned, most of the innovations in the BSF feed are the enrichment of SCG or other coffee parts with algae or their combination with various wastes. However, there is only one research study about the fermented SCG for BSF feed, as reported by Kharkratoke et al. [62]. They used fermented SCG in various proportions (0–100%) mixed with fruit and vegetable pulp residue for BSF feed. With 0% fermented SCG (or purely fruit and/or vegetable pulp residue), the waste reduction efficiency and feed conversion rate are the highest among all (~90%). With 20% fermented SCG mixed with fruit and vegetable pulp residue, the BSF larvae are developed in a short period of time (~15 days), with a high survival rate (~95%), and high bioconversion efficiency (~85%). The largest prepupal size is between 20% and 40% fermented SCG (similarly at around 120 mm length and 90 mg weight). However, those of 40% fermented SCG have a low substrate reduction rate (~65%) and a long prepupation time (~21 days), compared to those of 20% fermented SCG (~85% substrate reduction rate and a short prepupation time of 15 days). As the composition of fermented SCG is prolonged between 60–80%, the development time is getting longer to fall between 25–35 days, with a passable substrate reduction time of 55–68%. It is then worsened when BSF larvae are fed with 100% fermented SCG, as indicated by longer development times (45–53 days), a reduced larval survival rate (65%), and a substrate reduction rate (~30%).

An interesting result is reported in Table 1 by Horgan et al. [37]. They studied the effects of incorporating control SCG (aged < 1 months), aged SCG (7 and 14 months), and SCG-derived BSF frass into soil or 1 cm on soil on the growth of radish and tomato, while assessing their repellency to slug herbivory (*Arion atar, Deroceras laeve, Derocerus reticulatum,* and *Lehmannia marginata*). Fresh SCG inhibited plant growth while simultaneously reducing slug herbivory (radish plant height ~9 cm, radish leaf consumed by slug < 5%, tomato plant height ~7 cm, area of tomato leaf consumed by SCG < 1 mm^2). In addition, SCG-derived BSF frass mixed in the soil or on top of the soil resulted in declined plant growth, yellowing, and reduced height (radish plant height ~5 cm, tomato plant height 2.5 cm, where there was no further exploration for leaf consumption by slug). The 14-month-old SCG promoted plant growth with no repellency effect on slugs (radish plant height 13 cm, radish leaf consumed by slug 10–80%, tomato plant height 15 cm, area of tomato leaf consumed by SCG 10–35 mm^2). The best result is achieved when 7-month-old SCG is applied, where plant growth is promoted while simultaneously reducing slug herbivory (radish plant height 10 cm, radish leaf consumed by slug < 5%, tomato plant height 8 cm, area of tomato leaf consumed by SCG < 3 mm^2).

Related to the interesting effect of frass in the study by Horgan [37], BSF larvae frass is also stated as a biologically unstable soil amendment [63], which is a result of the rapid composting process in the presence of phytotoxic substances. In order to be suitable as a soil amendment, the BSF larvae frass have to be stabilized via post-treatment, for instance, thermophilic composting [63]. Ammonium nitrogen (NH_4^+-N) is highly suggested as the main driver of the phytotoxicity [64] in frass. This might be the reason why frass, as an organic excrement of BSF larvae, showed a negative effect on plant growth in the study by Horgan and coworkers [37].

3.2. Utilization of Coffee-Reared BSF Larvae in Fisheries

To explore the impacts of SCG on the growth of BSF larvae as a sustainable fish feed, Table 2 tabulates the particularly recent developments. Similar to Table 1, in Table 2, there is some leading research on the use of (1) coffee silverskin enriched with algae (*Schizochytrium* sp.) and (2) a mixture of SCG with other sources of nutrients (carbohydrate, protein, fibers). Those various feeds are then utilized to grow BSF larvae that will be further consumed by zebrafish (*Danio rerio*) or rainbow trout fish (*Oncorhynchus mykiss*). The effects observed in the fish fed with SCG-derived BSF larvae are detailed as follows:

Zarantoniello et al. [65] explored the use of coffee silverskin enriched with 10% *Schizochytrium* sp. algae to rear BSF larvae. The coffee silverskin-based BSF larvae are then used as a substitution for the zebrafish (*Danio rerio*) meal at various ratios. At 0% and 25% substitution rates of fish meal using coffee silverskin-based BSF larvae, the zebrafish show no significant change in their responses. This trend is continued up to the substitution level of 50%, which is the most balanced rate (in terms of fish growth and sustainability). Therefore, we can conclude that (to some extent) the BSF-based replacements provide comparable performance to the control fish meal diet. However, at higher levels of meal substitution (75 and 100%), negative effects are detected, such as severe hepatic steatosis, microbiota modification, increased lipid content, elevated stress, and immune responses. Based on the aforementioned results, their study demonstrates the potential of coffee silverskin-based BSF larvae as a promising partial substitution for fish meal, although there are some limits at high substitution levels.

Table 1. Compilation of the existing research of cultivation of BSF by using SCG as a feed component.

No.	Name of Rearing Substrate	Rearing Conditions	Results	Reference
1.	SCG	BSF larvae, 7 days old, feeding rate of 200, 100, 50, 25, and 12.5 mg/larvae/day	Best results for feeding with 200 mg/larvae/day – Development Time: 25.6 ± 2.19 days – Efficiency of digestion conversion: 5 ± 0.49% – Pupae Biomass: 14.8 ± 2.15 mg – Growth Rate: 1.41 ± 0.17 mg/day – Waste reduction index: 0.997 ± 0.11 – Pupae Protein Content: Up to 33.45%	[52]
2.	*Coffea arabica* (Castillo variety (0.5% caffeine))	BSF larvae, 100 mg/larvae/day	Reduction percentage of coffee pulp = 62.88% (wet basis) Efficiency of conversion of ingested food: 7.89%	[53]
		BSF larvae, 160 mg/larvae/day	Highest weight reduction index (wet basis) is 0.85%	
		BSF larvae, 200 mg/larvae/day	Highest weight = 115.9 mg. Shortest average development time = 38.65 days.	
3.	Coffee silverskin (100%) Coffee silverskin with algae (*Schizochytrium* sp., 0, 5, 10, 20, 25%) Coffee silverskin with algae (*Isochrysis galbana*, 0, 5, 10, 20, 25%)	BSF larvae of 6 days old, in 9 groups × 5 replicates × 150 larvae = 6750 larvae. Feeding rate = 100 mg/day. Chamber temperature = 27 ± 1 °C, relative humidity (RH) = 65 ± 5%, 24 h darkness in plastic boxes 28 × 19 × 14 cm^3. Note: Weight evolution of BSF larvae is not available	Best inclusion rate = *Schizochytrium* sp. 10%. The prepupae of BSF show increased lipid and protein content in prepupae, and high amounts of unsaturated fatty acids, especially omega-3. There is no information on the weight evolution of BSF. The BSF larvae were harvested when the tegument color changed from white to black.	[54]
4.	Coffee silverskin (100%) Coffee silverskin enriched with *Schizochytrium limacinum*, (5, 10, 20, 25%)	BSF larvae of 6 days old, in 9 groups × 5 replicates × 150 larvae = 6750 larvae. Feeding rate = 100 mg/day. Note: Weight evolution of BSF larvae is not available	High prevalence of tetracycline resistance genes. No significant effect on AR gene distribution in larvae.	[56]
	Coffee silverskin with *Isochrysis galbana* (5, 10, 20, 25%)		High prevalence of tetracycline resistance genes. Significant accumulation of AR genes in frass samples, especially at high percentages (>20%) of *I. galbana*.	

Table 1. *Cont.*

No.	Name of Rearing Substrate	Rearing Conditions	Results	Reference
5.	Coffee silverskin (100%)	BSF larvae of 6 days old, in 9 groups × 5 replicates × 150 larvae = 6750 larvae. Feeding rate = 100 mg/day. Chamber temperature = 27 ± 1 °C, 65 ± 5% RH, 24 h dark photoperiod, in plastic boxes 28 × 19 × 14 cm^3. Note: Weight evolution of BSF larvae is not available	Dominance of *Paenibacillus* in the larvae's microbiota	[57]
	Coffee silverskin enriched with *Schizochytrium limacinum*, (5, 10, 20, 25%)		Presence of *Enterococcus*, *Lysinibacillus*, *Morganella*, and *Paenibacillus* in the larvae's microbiota. Dominance of *Brevundimonas* and *Alcaligenes* in frass.	
	Coffee silverskin with *Isochrysis galbana*, (5, 10, 20, 25%)		High relative abundances of *Brevundimonas*, *Enterococcus*, *Paracoccus*, and *Paenibacillus* in larvae. Predominance of *Brevundimonas* in frass.	
6.	Brewery spent grains, tomato peels and seeds, cows' milk, whey, grape stalks, bread dough, and SCG.	Chamber (32.5 × 32.5 × 32.5 cm^3) temperature = 27 ± 0.5 °C, 60–70% RH, with light:dark = 16:8 h, with 400–500 larvae per chamber.	– Best larval performance at 0.22 g/prepupae (beer), 0.19 g/prepupae (tomato peels), and 0.14 g/prepupae (cheese). – No development for SCG-fed BSF larvae (died after 15 days, survival rate = 0%). – Highest ratio between unsaturated and saturated fatty acids at 1.34 (cow's milk and whey), while the lowest ratio is 0.37 (tomato peels and seeds).	[59]
7.	SCG	BSF larvae and prepupae fed for 35 days. Feeding of 6.8 kg feed in container 0.9 × 1.2 × 1.5 m^3.	Survival: 45%. Longer and heavier BSFP from dough. Fatty acids higher in BSFL. Frass: Higher potassium (~1.00% dry matter (DM)), lowest phosphorus (<0.30% DM), moderate nitrogen content (~3.25% DM) Larval length = 16.86 ± 0.29 mm Larval weight = 0.11 ± 0.01 g Net production = 0.75 ± 0.58 g/day/m^3	[45]
	Donut dough		Survival: 24%. Longer and heavier BSFP from dough. Lowest nitrogen content (~2.75% DM), lowest potassium (<0.25% DM), lowest calcium (<0.1% DM). Higher amino acid composition in BSFL. Larval length = 21.44 ± 0.59 mm Larval weight = 0.23 ± 0.01 g Net production = 0.60 ± 1.01 g/day/m^3	
	Blend (coffee and dough, 1:1)		Survival: 81%. Stage and food affected protein, lipid, glycogen content. Frass: Highest nitrogen (~4.20% DM). Comparable to soybean meal and organic fertilizers. Larval length = 19.12 ± 0.72 mm Larval weight = 0.18 ± 0.01 g Net production = 4.42 ± 1.02 g/day/m^3	

Table 1. *Cont.*

No.	Name of Rearing Substrate	Rearing Conditions	Results	Reference
	SCG		Larvae underperformed. Development time = 35 days. Larval yield 0.61–0.77 g in DM.	
	Brewer's spent grain (BSG)		Better results than those solely reared with SCG. Development time = 9 days. Larval yield 6.02–8.53 g in DM (for 200 and 300 larvae per container, respectively).	
8.	SCG and/or BSG and/or brewers' yeast (BY)	BSF rearing with controlled climate chamber (14.5 × 9.5 × 9.5 cm^3, 30 °C, 70% RH, with 24 h dark photoperiod. Total amount of feed per larva = 240 mg in DM. Variation: 200 and 300 larvae per container (1.45 and 2.17 larvae per cm^2, respectively.	Acceptable values in substrate mass reduction (18.48–44.80%, SCG + BY and BSG + BY, respectively), protein conversion rate (17.55–30.15%, SCG + BY and BSG + BY, respectively), and bioconversion rate (6.25–17.22%, SCG + BY and BSG + BY, respectively). Development time = 8–17 days (for BSG + BY, and SCG + BY, respectively). Larval yields 2.99–12.72 g in DM (for SCG + BY 200 and BSG + BY 300, respectively). For SCG + BSG + BY, the development time is 12 days, with larval yield of 6.93–10.53 g in DM (for 200 and 300 larvae, respectively).	[61]
	Reference feed (layer feed BΩ-321 (Viozois S.A., Athens, Greece))		Highest values in substrate mass reduction (67.38%), protein conversion rate (45.77%), and bioconversion rate (23.28%) among all feeds. Development time = 11 days. Larval yield 11.32–16.78 g in DM (for 200 and 300 larvae, respectively).	

Table 1. *Cont.*

No.	Name of Rearing Substrate	Rearing Conditions	Results	Reference
9.	Mixture of sludge-containing media (S), with brewery spent grains (BSG), coffee waste (C), and whey (W).	Fed to BSF with composition of S:BSG:C:W = 50:10:10:30, reared for 23 days.	Larvae did not complete their development, with very poor performance of: Final height = ~6 mm Final weight ≤ 10 mg	[46]
		Control (feed from Entocycle.com, with unknown composition)	Final height = ~15–18 mm Final weight = 94.5 ± 7.2 mg	
		S: BSG:C:W = 50:10:30:10	Final height = ~15–18 mm Final weight = 81.9 ± 5.3 mg	
		S: BSG:C:W = 50:50:0:0	Final height = ~15–18 mm Final weight = 91.0 ± 0.1 mg	
		S: BSG:C:W = 50:30:10:10	Final height = ~15–18 mm Final weight = 84.0 ± 6.0 mg	
10.	Sweet potato, 2 weeks	BSF larvae cultivated for 2 weeks. Temperature controlled room of 3.6 × 3.6 m^2, 30 °C, 60–64% RH.	– Material reduction = 29.9 ± 1.3% – Percentage of prepupae = 36.4 ± 5.6% – Survival rate = 87.0 ± 5.6% – Length (larvae, prepupae) = ~16.8 mm, ~17.6 mm – Weight (larvae, prepupae) = ~16.9 mg, ~17.5 mg – Highest percentage (~43%) of lauric acid (C12) – Supported growth and best nutritional profile of BSF larvae – High total amino acid content (38.33 ± 2.39%)	[47]
	SCG, 2 weeks		– Material reduction = 63.4 ± 3.3% – Percentage of prepupae = 36.5 ± 5.8% – Survival rate = 98.6 ± 1.3% – Length (larvae, prepupae) = ~16.7 mm, ~17.4 mm – Weight (larvae, prepupae) = ~16.8 mg, ~17.2 mg – Significantly higher crude protein than the other diets (~65% dry weight, while others are ~40%) – Highest composition of C16:0 (palmitic acid, 24.85%) and C18:2n-6 (linoleic acid, 23.30%) compared to those fed with sweet potato (17.45%, 12.40%, respectively) or dough (15.21%, 18.92%, respectively). – Highest total amino acid content (39.49 ± 2.57%)	

Table 1. *Cont.*

No.	Name of Rearing Substrate	Rearing Conditions	Results	Reference
	Dough, 2 weeks		– Material reduction = 92.6 ± 12.2% (best material reduction) – Percentage of prepupae = 15.1 ± 1.2% (lowest percentage of prepupae among sweet potatoes and dough). – Survival rate = 83.3 ± 2.9% – Length (larvae, prepupae) = ~18.5 mm, ~19.5 mm (significantly larger/heavier BSF larvae). – Weight (larvae, prepupae) = ~18.5 mg, ~19.5 mg – Lowest total amino acid content (30.33 ± 0.18%)	
11.	Fruit and vegetable pulp residue (apple 12%, pineapple 12%, carrot 25%, tomato 44%, guava 5%, beetroot 1.5%, celery 0.5%) mixed with 0% fermented SCG	Larvae kept in plastic containers with diameter 24.5 cm × height 12.5 cm, with 28–34 °C, 70–90% RH, light:dark = 13 h: 11 h. Feeding rate = 200 mg/larvae/day.	– Survival rate = ~85% – Duration of prepupation = ~10 days – Weight (larvae, prepupae) = 121 mg, 64 mg – Substrate reduction rate = ~90% – Highest waste reduction efficiency and feed conversion rate, similar to 20% fermented SCG. Served as the control.	[62]
	Pulp with 20% fermented SCG		– Survival rate = ~95% – Duration of prepupation = ~15 days – Weight (larvae, prepupae) = 116 mg, 90 mg – Substrate reduction rate = ~85% – Short rearing time, high survival rate, and high substrate reduction rate. Supported the largest prepupae size. – Recommended for rearing BSF larvae.	
	Pulp with 40% fermented SCG		– Survival rate = ~92% – Duration of prepupation = ~21 days – Weight (larvae, prepupae) = 136 mg, 91 mg – Substrate reduction rate = ~65% – Largest prepupae size, high survival rate, but with long prepupation duration and not optimum reduction rate.	

Table 1. *Cont.*

No.	Name of Rearing Substrate	Rearing Conditions	Results	Reference
	Pulp with 60% fermented SCG		– Duration of prepupation = 25–35 days – Weight (larvae, prepupae) = 139 mg, 81 mg – Substrate reduction rate = ~68% – Large prepupae size, passable reduction rate, but long prepupation duration.	
	Pulp with 80% fermented SCG		– Survival rate = ~80% – Duration of prepupation = 30–35 days (long development time) – Weight (larvae, prepupae) = 102 mg, 79 mg – Substrate reduction rate = ~55%	
	Pulp with 100% fermented SCG		– Survival rate = ~65% – Duration of prepupation = 45–53 days – Weight (larvae, prepupae) = 117 mg, 74 mg – Substrate reduction rate = ~30% – Very long development time, low rate of substrate reduction, lightest prepupae weight.	
12.	Fresh SCG (aged < 1 month, 50% water, incorporated into soil)	BSF larvae grown in the first chamber (275 × 250 × 210 cm^3, 25–28 °C, 60% RH) for 3–5 days. After that, they will be grown in the second chamber (275 × 80 × 250 cm^3, 28 ± 1.5 °C, 40% RH, 24 h dark photoperiod) for 12 days. Growing of radish and tomato = 30 days	Inhibited plant growth (radish and tomato, 30 days) and development; reduced slug herbivory (*Arion atar*, *Deroceras laeve*, *Derocerus reticulatum*, and *Lehmannia marginata*). Height of radish plant = ~9 cm Radish leaf consumed by slug ≤ 5% Height of tomato plant = ~7 cm Area of tomato leaf consumed by slug (SCG mixed with soil or layered on top of the soil) = 0 mm^2, <1 mm^2	[37]
	SCG (aged 7 months, 1 cm top dressing)		Promoted growth, while simultaneously reduced slug herbivory through repellent and host quality effects. Height of radish plant = ~10 cm Radish leaf consumed by slug ≤ 5% Height of tomato plant = ~8 cm Area of tomato leaf consumed by slug (SCG mixed with soil or layered on top of the soil) = 0 mm^2, <3 mm^2	

Table 1. *Cont.*

No.	Name of Rearing Substrate	Rearing Conditions	Results	Reference
	SCG (aged 14 months, incorporated into soil)		Promoted plant growth, while having no effect on slug herbivory. Height of radish plant = ~13 cm Radish leaf consumed by slug (SCG mixed with soil, or layered on top of the soil) = 10–80% Height of tomato plant = ~15 cm Area of tomato leaf consumed by slug (SCG mixed with soil or layered on top of the soil) = 10 mm^2, 35 mm^2	
	SCG-derived BSF frass (incorporated into soil, or 1 cm top dressing)		Reduced development of plants, yellowing, reduced height. Height of radish plant = ~5 cm Radish leaf consumed by slug (SCG mixed with soil or layered on top of the soil) = not available Height of tomato plant = ~2.5 cm Area of tomato leaf consumed by slug (SCG mixed with soil or layered on top of the soil) = not available	

On yet another use of coffee silverskin enriched with *Schizochytrium* sp. microalgae for rearing BSF larvae, Milanović et al. explored the dynamic profile of the antibiotic resistance (AR) genes in zebrafishes that are fed with them [66]. The study employed quantitative Polymerase Chain Reaction (qPCR) to assess the detection of various AR genes. It is shown by the researchers that genes like erm(B), tet(K), tet(M), tet(O), and tet(S) are present. On the other hand, mecA, vanA, vanB, and aac-aph genes are not detected. Amusingly, in zebrafish (at both juvenile and reproductive stages), the AR genes such as erm(A), erm(C), vanB, and aac-aph are never detected, whereas erm(B), tet(M), and tet(S) are widespread.

Rodrigues et al. investigated the use of fish discards and coffee silverskin enriched with *Schizochytrium* sp. for rearing BSF larvae [67]. It was found that those substrates significantly improved the lipid profile of BSF larvae, enhancing the content of omega-3 fatty acids, which is crucial nutrition for fish health (and also human health). Upon using fish discards, there is approximately a 40% improvement in the lipid profile and essential fatty acid profile. In addition, the practice of combining fish discards with coffee silverskin enriched with *Schizochytrium* sp. leads to high levels of eicosapentaenoic acid (EPA) and docosahexaenoic acid (DHA) detection in BSF larvae.

Chemello et al. assessed the physiological responses of zebrafish fed with BSF larvae reared on coffee silverskin enriched with 10% *Schizochytrium* sp. algae [48]. They were then used as a substitute for fish meal at different levels (0, 20, 50, 75, and 100%). At lower replacement levels (0, 20, and 50%), there are no physiological issues observed in the zebrafish, which is an indication that the larvae are a viable component of the diet. However, at higher substitution levels (75 and 100%), there are significant impacts on fish stress response, oocyte maturation stages, spawning, and hatching success. This study provides critical insights into the feasibility of using BSF larvae as a fish meal substitute, highlighting the importance of moderation in substitution levels to avoid adverse effects on fish health.

The study by Osimani et al. explored the effects of feeding zebrafish with BSF larvae and BSF frass reared on coffee by- products [68]. Variability in viable bacterial counts based on the substrate used are shown in the zebrafish. When they are fed with larvae reared on coffee by-products, *Danio rerio* exhibited a higher occurrence of *Lactobacillaceae* and *Leuconostoccaceae*, low bacterial richness and evenness, and a higher abundance of *Clostridia*, influenced by bioactive compounds from the coffee by-products. As an interesting finding, the core fish microbiota remained stable despite these changes. On the other hand, the zebrafishes fed with larvae that are reared on a mixture of vegetables showed the highest bacterial richness and evenness in larvae and frass. They also showed higher abundance of *Enterobacteriaceae*, as well as a stable core microbiota.

In a recent study by Ratti et al., *Oncorhynchus mykiss* (rainbow trout) were fed with different proportions of BSF meal for six weeks [49]. The control group receives a regular BSF meal, while the test groups are fed 3% and 20% BSF prepupae meals. The findings reveal that the fishes accept the BSF-based diet well, with no adverse effects on fish growth, gut and liver health, or marketable characteristics. Interestingly, the group fed with 20% BSF prepupae meal demonstrated an increase in immuno-related gene expression, with a slight reduction in fillet redness and yellowness.

Table 2. Recent studies on the effects of coffee-reared BSF larvae as feed for fishe.

No.	Name of Rearing Substrate	Rearing Conditions	Produced BSF Fed to	Results	Reference
1.	BSF reared on coffee silverskin enriched with 10% *Schizochytrium* sp.	BSF as substitution of fish meal (0, 25, 50, and 75%)	Zebrafish (*Danio rerio*)	At 0% (baseline) and 25%, zebrafishes show standard responses based on the control diet. The 50% substitution gives the best compromise between sustainability and proper fish growth. However, at 75% and 100% substitution, severe hepatic steatosis was observed, along with microbiota modification, increased lipid content, fatty acid modification, and higher expression of stress and immune response markers.	[65]
2.	BSF reared on coffee silverskin with 10% *Schizochytrium* sp. microalgae	BSF as substitution of fish meal (0, 25, 50, 75, and 100%) for 6 months	Zebrafish (*Danio rerio*)	This study employed qPCR to assess the dynamics of antibiotic resistance (AR) genes in fish feed, including those with insect meal ingredients. Resistant genes studied: macrolide-lincosamide-streptogramin B (MLSB) [erm(A), erm(B), erm(C)], vancomycin (vanA, vanB), tetracyclines [tet(M), tet(O), tet(S), tet(K)], β-lactams (mecA, blaZ), and aminoglycosides [aac-aph]. Findings in diet samples: Detected: erm(B), tet(K), tet(M), tet(O), and tet(S). Not Detected: mecA, vanA, vanB, and aac-aph. Findings in zebrafish (juvenile and reproductive stages): Never Detected: erm(A), erm(C), vanB, and aac-aph. Widespread: erm(B), tet(M), and tet(S).	[66]
		0% BSF larvae		erm(A), erm(C), vanB, aac-aph not detected in zebrafish at any stage.	
3.	Marine-based substrates		Zebrafish	– Significantly improves the lipid profile of BSF larvae. – Enhanced omega-3 fatty acids profile.	[67]
	Fish discards			– Improvement of approximately 40% in the lipid profile. – Increase in essential fatty acids.	
	Fish discards with coffee silverskin with *Schyzochytrium* sp.			– High levels of EPA and DHA recorded. – Still reliant on marine-based bioresources	
	Coffee silverskin enriched with 10% of *Schizochytrium* sp.	100% fish meal replacement		Affected fish stress response, oocytes maturation stages, spawning, and hatching success.	

Table 2. *Cont.*

No.	Name of Rearing Substrate	Rearing Conditions	Produced BSF Fed to	Results	Reference
4.	Coffee silverskin enriched with 10% of *Schizochytrium* sp.	0% fish meal replacement	Zebrafish	No impairment observed in zebrafish physiological responses	[48]
		20% fish meal replacement			
		50% fish meal replacement			
		75% fish meal replacement		Affected fish stress response, oocytes maturation stages, spawning, and hatching success.	
		100% fish meal replacement			
5.	Coffee by-products	BSF larvae and frass	Zebrafish (*Danio rerio*)	Zebrafishes fed with BSF larvae reared on coffee by products show: – Variability in viable counts based on substrate. – Highest occurrence of *Lactobacillaceae* and *Leuconostoccaceae*. – Low bacterial richness/evenness. – Higher abundance of *Clostridia* in zebrafish influenced by bioactive compounds from coffee by-products. – Core fish microbiota remained stable.	[68]
	Mixture of vegetables			Zebrafishes fed by BSF larvae reared on mixture of vegetable show: – Highest bacterial richness/evenness in larvae and frass of BSF. – Low bacterial richness/evenness. – Higher abundance of *Enterobacteriaceae* in zebrafish, possibly influenced by bioactive compounds (chlorogenic and caffeic acids) in the substrate. – Core fish microbiota remained stable.	
6.	BSF meal (control)	Fed 6 weeks	*Oncorhynchus mykiss*	Well accepted; no impairment in fish growth, gut and liver health, or marketable characteristics.	[49]
	3% BSF prepupae meal	Fed 6 weeks			
	20% BSF prepupae meal	Fed 6 weeks		Well accepted; increased immuno-related gene expression and slight reduction of fillet redness and yellowness.	

3.3. Bibliometric Analysis of the Effects of Feeding SCG to BSF

In order to illustrate the landscape of research on feeding SCG to the cultivation of BSF, a bibliometric analysis is conducted. The author keywords from the 18 core papers are entangled in an orderly fashion, as shown in Figure 2, as a result of the VOSViewer software (version 1.6.17). There are four clusters that can be observed, as follows:

1. BSF reared with SCG for aquaculture (red)
2. (Bio)chemical analysis related to BSF and SCG (green)
3. BSF and bioconversion (blue)
4. SCG and other substrates (yellow)

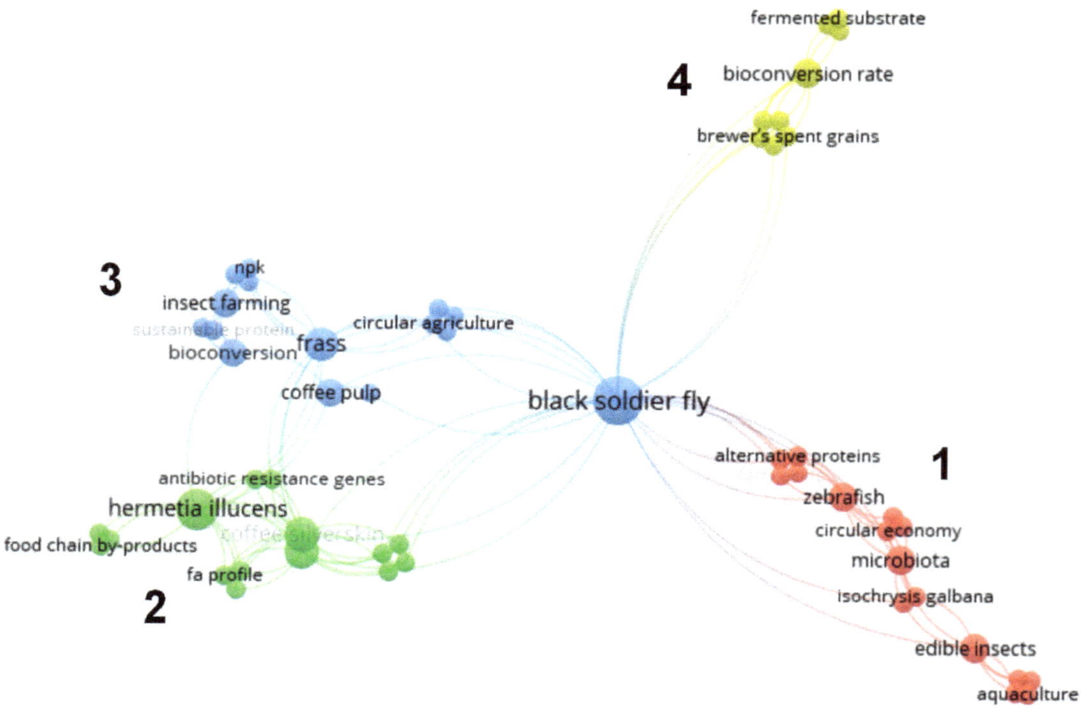

Figure 2. The bibliometric visualization of the research on the effects of SCG to the BSF cultivation.

The details of the members of the aforementioned clusters and their strength (weight) are detailed in Table 3.

The mapping of the keywords in Figure 2 resulted from the bibliometric analysis of the keywords generated by the authors ("author keywords") of the 18 core papers on the usage of SCG or related coffee parts for the rearing of BSF larvae. There are 55 keywords that are correlated with each other, with 162 links and a total link strength of 166. They are clustered together, with the minimum cluster size optimized to be 8.

Table 3. All keywords from the 18 key papers in this study, as visualized in Figure 2.

Cluster 1	Weight	Cluster 2	Weight	Cluster 3	Weight	Cluster 4	Weight
zebrafish	9	coffee silverskin	13	black soldier fly	28	bioconversion rate	9
edible insects	8	*Hermetia illucens*	13	frass	14	brewer's spent grains	6
microbiota	8	microalgae	13	insect farming	7	protein conversion	6
alternative proteins	5	bioaccumulation	6	bioconversion	5	spent coffee grounds	6
fatty acids	5	chemical hazard	6	circular agriculture	5	substrate mass reduction	6
reproduction	5	*Hermetia illucens* prepupae	6	integrated pest management	5	upcycling	6
Schizochytrium sp.	5	potentially toxic elements	6	repellence	5	fermented substrate	3
aquaculture	4	antibiotic resistance genes	5	systemic defenses	5	growth performances	3
circular economy	4	fa profile	5	coffee pulp	4	SCG	3
fish feed	4	principal component analysis	5	npk	4		
insect meal	4	rearing substrates	5	prepupae	4		
Isochrysis galbana	4	relative macromolecular composition	5	spent coffee	4		
polyunsaturated fatty acids	4	food chain by-products	3	sustainable protein	3		
qPCR	4	prepupal fatty acids profile	3	waste management	3		
Schizochytrium limacinum	4	waste valorization	3	caffeine	2		
tetracyclines	4						

3.4. Bibliometric Analysis of a Single Keyword: "Black Soldier Fly"

The bibliometric analysis conducted via VOSViewer software allows us to specifically assess the multiple connections of one keyword of interest. That particular keyword is "spent coffee ground". This is quite a niche keyword for BSF cultivation that might have plenty of room for innovation in the future. Based on the data in Table 3, the keyword "spent coffee ground" is placed in cluster 4 (yellow), with a weight of 6. The connections to the single keyword "spent coffee ground" based on VOSViewer analysis for the 18 core papers are shown in Figure 3 and detailed in Table 4. Please kindly note that a gray ellipse is intentionally placed to cover most of the keywords not related to the connection stemming from the keyword "spent coffee ground".

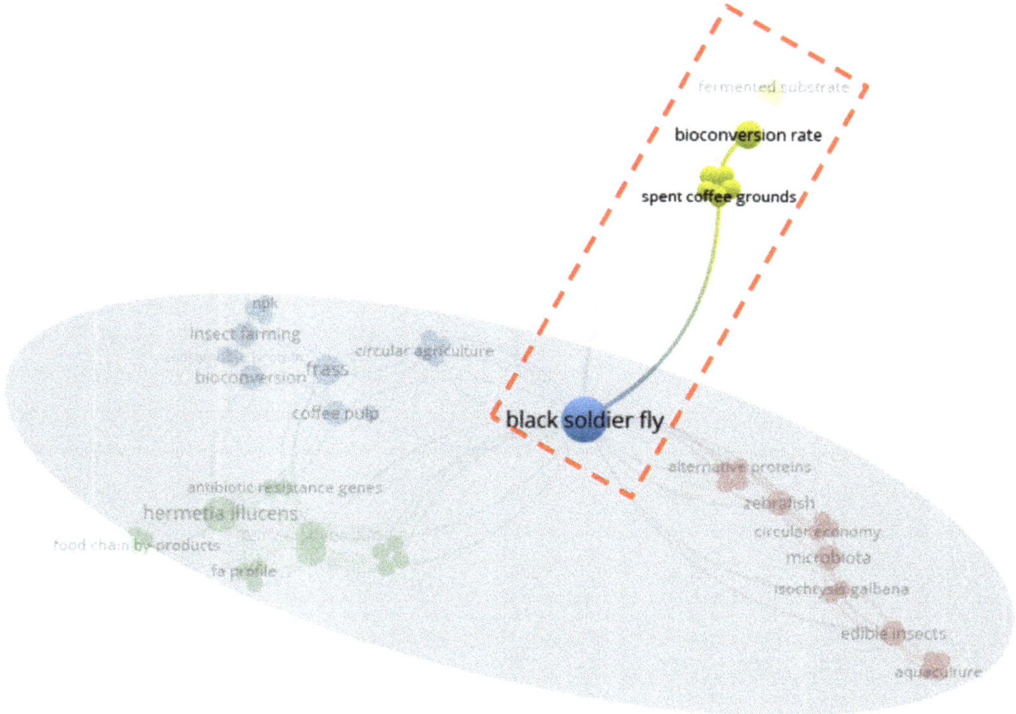

Figure 3. Analysis of keyword "spent coffee ground" as a particular interest. Links = 6, total link strength = 6.

Table 4. Network analysis for the terms associated with the keyword "spent coffee ground" obtained from Figure 3.

No.	First Node	Second Node	Cluster for the First Node	Cluster for the Second Node	Link Strength
1.	spent coffee grounds	substrate mass reduction	4	4	1
2.	spent coffee grounds	upcycling	4	4	1
3.	black soldier fly	spent coffee grounds	3	4	1
4.	bioconversion rate	spent coffee grounds	4	4	1
5.	brewer's spent grains	spent coffee grounds	4	4	1
6.	protein conversion	spent coffee grounds	4	4	1

Based on Figure 3 and Table 4, it can be concluded that spent coffee grounds in the cultivation of BSF larvae are clearly connected with the keyword "black soldier fly", which enables the "substrate mass reduction" with a specific "bioconversion rate" for the beneficial process of "upcycling" organic wastes such as "brewer's spent grains" (among other resources) to obtain "protein conversion". The portion highlighted in pale yellow in Figure 3 is the area of the gap in the research on BSF cultivation using SCG and/or other coffee parts. It is clearly seen that there is indeed plenty of room for improvement and innovation in this research area. To name a few interesting topics, there are "food chain by-products", "circular agriculture", "antibiotic resistance genes", "alternative proteins", "aquaculture", and "circular economy".

4. Conclusions

This study reports the recent development of the utilization of SCG (and/or coffee parts, mixed with or without other organic materials) for the cultivation of BSF larvae. As coffee contains caffeine as an alkaloid that hinders the growth of some plants or the BSF larvae themselves, the right composition or modification of SCG is intriguing to be explored. The recent usage of SCG-derived or coffee-fed BSF larvae in aquaculture and fisheries is also reported here, along with the limitations of substitution levels and the related consequences for the fishes. Bibliometric analyses are also performed to supplement the mapping of the recent advances in the cultivation and application of BSF larvae using SCG and/or coffee parts.

Author Contributions: Conceptualization: D.M.; Funding acquisition: D.M.; Investigation: D.J.C.H.; Methodology: D.J.C.H. and D.M.; Resources: D.J.C.H.; Validation: D.J.C.H.; Visualization: D.J.C.H.; Writing—original draft: D.J.C.H. and D.M.; Writing—review and editing: D.J.C.H. and D.M. All authors have read and agreed to the published version of the manuscript.

Funding: This work was supported by the Directorate of Resources, Directorate General of Higher Education, Ministry of Education, Culture, Research, and Technology, Republic of Indonesia, as a part of Penelitian Dasar Unggulan Perguruan Tinggi (PDUPT) to Bina Nusantara University with contract numbers: 073/E5/PG.02.00.PL/2023; 1165/LL3/AL.04/2023; 095/VR.RTT/V/2023.

Institutional Review Board Statement: Not applicable.

Data Availability Statement: The data that support the findings of this study are available from the corresponding author (D.M.) upon reasonable request.

Acknowledgments: We would like to thank the members of the multidisciplinary research group of the Waste-Food-Environmental Nexus Research Interest Group of Bina Nusantara University for their academic and non-academic support.

Conflicts of Interest: The authors declare no conflicts of interest.

References

1. Hansen, R.J.; Nielsen, S.H.M.; Johansen, M.; Nielsen, F.K.N.; Dragsbæk, F.B.; Sørensen, O.S.B.; Eriksen, N.T. Metabolic performance of black soldier fly larvae during entomoremediation of brewery waste. *J. Appl. Entomol.* **2023**, *147*, 423–431. [CrossRef]
2. Bekker, N.S.; Heidelbach, S.; Vestergaard, S.Z.; Nielsen, M.E.; Riisgaard-Jensen, M.; Zeuner, E.J.; Bahrndorff, S.; Eriksen, N.T. Impact of substrate moisture content on growth and metabolic performance of black soldier fly larvae. *Waste Manag.* **2021**, *127*, 73–79. [CrossRef] [PubMed]
3. Eriksen, N.T. Dynamic modelling of feed assimilation, growth, lipid accumulation, and CO_2 production in black soldier fly larvae. *PLoS ONE* **2022**, *17*, e0276605. [CrossRef]
4. Laganaro, M.; Bahrndorff, S.; Eriksen, N.T. Growth and metabolic performance of black soldier fly larvae grown on low and high-quality substrates. *Waste Manag.* **2021**, *121*, 198–205. [CrossRef]
5. Mangindaan, D.; Kaburuan, E.; Meindrawan, B. Black Soldier Fly Larvae (*Hermetia illucens*) for Biodiesel and/or Animal Feed as a Solution for Waste-Food-Energy Nexus: Bibliometric Analysis. *Sustainability* **2022**, *14*, 13993. [CrossRef]
6. Banks, I.J.; Gibson, W.T.; Cameron, M.M. Growth rates of black soldier fly larvae fed on fresh human faeces and their implication for improving sanitation. *Trop. Med. Int. Health* **2014**, *19*, 14–22. [CrossRef]
7. Nguyen, T.T.X.; Tomberlin, J.K.; Vanlaerhoven, S. Ability of Black Soldier Fly (Diptera: Stratiomyidae) Larvae to Recycle Food Waste. *Environ. Entomol.* **2015**, *44*, 406–410. [CrossRef]
8. Diener, S.; Zurbrügg, C.; Tockner, K. Conversion of organic material by black soldier fly larvae: Establishing optimal feeding rates. *Waste Manag. Res.* **2009**, *27*, 603–610. [CrossRef]
9. Lalander, C.; Diener, S.; Magri, M.E.; Zurbrügg, C.; Lindström, A.; Vinnerås, B. Faecal sludge management with the larvae of the black soldier fly (*Hermetia illucens*)—From a hygiene aspect. *Sci. Total Environ.* **2013**, *458–460*, 312–318. [CrossRef] [PubMed]
10. Makkar, H.P.S.; Tran, G.; Heuzé, V.; Ankers, P. State-of-the-art on use of insects as animal feed. *Anim. Feed Sci. Technol.* **2014**, *197*, 1–33. [CrossRef]
11. Spranghers, T.; Ottoboni, M.; Klootwijk, C.; Ovyn, A.; Deboosere, S.; De Meulenaer, B.; Michiels, J.; Eeckhout, M.; De Clercq, P.; De Smet, S. Nutritional composition of black soldier fly (*Hermetia illucens*) prepupae reared on different organic waste substrates. *J. Sci. Food Agric.* **2017**, *97*, 2594–2600. [CrossRef]
12. Barroso, F.G.; de Haro, C.; Sanchez-Muros, M.-J.; Venegas, E.; Martinez-Sanchez, A.; Pérez-Bañón, C. The potential of various insect species for use as food for fish. *Aquaculture* **2014**, *422–423*, 193–201. [CrossRef]

13. Renna, M.; Schiavone, A.; Gai, F.; Dabbou, S.; Lussiana, C.; Malfatto, V.; Prearo, M.; Capucchio, M.T.; Biasato, I.; Biasibetti, E.; et al. Evaluation of the suitability of a partially defatted black soldier fly (*Hermetia illucens* L.) larvae meal as ingredient for rainbow trout (*Oncorhynchus mykiss* Walbaum) diets. *J. Anim. Sci. Biotechnol.* **2017**, *8*, 57. [CrossRef] [PubMed]
14. Elhag, O.; Zhou, D.; Song, Q.; Soomro, A.A.; Cai, M.; Zheng, L.; Yu, Z.; Zhang, J. Screening, expression, purification and functional characterization of novel antimicrobial peptide genes from *Hermetia illucens* (L.). *PLoS ONE* **2017**, *12*, e0169582. [CrossRef] [PubMed]
15. Badwan, A.A.; Rashid, I.; Al Omari, M.M.H.; Darras, F.H. Chitin and Chitosan as Direct Compression Excipients in Pharmaceutical Applications. *Mar. Drugs* **2015**, *13*, 1519–1547. [CrossRef]
16. Sheppard, D.C.; Tomberlin, J.K.; Joyce, J.A.; Kiser, B.C.; Sumner, S.M. Rearing methods for the black soldier fly (diptera: Stratiomyidae). *J. Med. Entomol.* **2002**, *39*, 695–698. [CrossRef]
17. Rummel, P.S.; Beule, L.; Hemkemeyer, M.; Schwalb, S.A.; Wichern, F. Black Soldier Fly Diet Impacts Soil Greenhouse Gas Emissions from Frass Applied as Fertilizer. *Front. Sustain. Food Syst.* **2021**, *5*, 709993. [CrossRef]
18. Rehman, K.U.; Cai, M.; Xiao, X.; Zheng, L.; Wang, H.; Soomro, A.A.; Zhou, Y.; Li, W.; Yu, Z.; Zhang, J. Cellulose decomposition and larval biomass production from the co-digestion of dairy manure and chicken manure by mini-livestock (*Hermetia illucens* L.). *J. Environ. Manag.* **2017**, *196*, 458–465. [CrossRef]
19. Oonincx, D.G.A.B.; de Boer, I.J.M. Environmental Impact of the Production of Mealworms as a Protein Source for Humans—A Life Cycle Assessment. *PLoS ONE* **2012**, *7*, e51145. [CrossRef]
20. Gold, M.; Tomberlin, J.K.; Diener, S.; Zurbrügg, C.; Mathys, A. Decomposition of biowaste macronutrients, microbes, and chemicals in black soldier fly larval treatment: A review. *Waste Manag.* **2018**, *82*, 302–318. [CrossRef]
21. Li, Q.; Zheng, L.; Cai, H.; Garza, E.; Yu, Z.; Zhou, S. From organic waste to biodiesel: Black soldier fly, *Hermetia illucens*, makes it feasible. *Fuel* **2011**, *90*, 1545–1548. [CrossRef]
22. Campos-Vega, R.; Loarca-Pina, G.; Vergara-Castaneda, H.A.; Oomah, B.D. Spent coffee grounds: A review on current research and future prospects. *Trends Food Sci. Technol.* **2015**, *45*, 24–36. [CrossRef]
23. Chen, X.-E.; Mangindaan, D.; Chien, H.-W. Green sustainable photothermal materials by spent coffee grounds. *J. Taiwan Inst. Chem. Eng.* **2022**, *137*, 104259. [CrossRef]
24. Chien, H.-W.; Kuo, C.-J.; Kao, L.-H.; Lin, G.-Y.; Chen, P.-Y. Polysaccharidic spent coffee grounds for silver nanoparticle immobilization as a green and highly efficient biocide. *Int. J. Biol. Macromol.* **2019**, *140*, 168–176. [CrossRef] [PubMed]
25. Hsieh, P.-C.; Chen, Y.-C.; Zheng, N.-C.; Mangindaan, D.; Chien, H.-W. A low-cost and environmentally-friendly chitosan/spent coffee grounds composite with high photothermal properties for interfacial water evaporation. *J. Ind. Eng. Chem.* **2023**, *126*, 283–291. [CrossRef]
26. Mangindaan, D.; Lin, G.-Y.; Kuo, C.-J.; Chien, H.-W. Biosynthesis of silver nanoparticles as catalyst by spent coffee ground/recycled poly(ethylene terephthalate) composites. *Food Bioprod. Process.* **2020**, *121*, 193–201. [CrossRef]
27. Ballesteros, L.F.; Teixeira, J.A.; Mussatto, S.I. Chemical, Functional, and Structural Properties of Spent Coffee Grounds and Coffee Silverskin. *Food Bioprocess Technol.* **2014**, *7*, 3493–3503. [CrossRef]
28. Murthy, P.S.; Naidu, M.M. Recovery of Phenolic Antioxidants and Functional Compounds from Coffee Industry By-Products. *Food Bioprocess Technol.* **2012**, *5*, 897–903. [CrossRef]
29. Kondamudi, N.; Mohapatra, S.K.; Misra, M. Spent Coffee Grounds as a Versatile Source of Green Energy. *J. Agric. Food Chem.* **2008**, *56*, 11757–11760. [CrossRef]
30. Caetano, N.S.; Silva, V.F.M.; Melo, A.C.; Martins, A.A.; Mata, T.M. Spent coffee grounds for biodiesel production and other applications. *Clean Technol. Environ. Policy* **2014**, *16*, 1423–1430. [CrossRef]
31. Liu, Y.; Tu, Q.; Knothe, G.; Lu, M. Direct transesterification of spent coffee grounds for biodiesel production. *Fuel* **2017**, *199*, 157–161. [CrossRef]
32. Tyagi, V.K.; Fdez-Güelfo, L.A.; Zhou, Y.; Álvarez-Gallego, C.J.; Romero Garcia, L.I.; Ng, W.J. Anaerobic co-digestion of organic fraction of municipal solid waste (OFMSW): Progress and challenges. *Renew. Sustain. Energy Rev.* **2018**, *93*, 380–399. [CrossRef]
33. Ayodele, O.O.; Adekunle, A.E.; Adesina, A.O.; Pourianejad, S.; Zentner, A.; Dornack, C. Stabilization of anaerobic co-digestion of biowaste using activated carbon of coffee ground biomass. *Bioresour. Technol.* **2021**, *319*, 124247. [CrossRef] [PubMed]
34. Iriondo-DeHond, A.; Rios, M.B.; Herrera, T.; Rodriguez-Bertos, A.; Nuñez, F.; San Andres, M.I.; Sanchez-Fortun, S.; Del Castillo, M.D. Coffee Silverskin Extract: Nutritional Value, Safety and Effect on Key Biological Functions. *Nutrients* **2019**, *7*, 2693. [CrossRef] [PubMed]
35. Mitraka, G.-C.; Kontogiannopoulos, K.N.; Batsioula, M.; Banias, G.F.; Assimopoulou, A.N. Spent Coffee Grounds' Valorization towards the Recovery of Caffeine and Chlorogenic Acid: A Response Surface Methodology Approach. *Sustainability* **2021**, *13*, 8818. [CrossRef]
36. Borrelli, R.C.; Esposito, F.; Napolitano, A.; Ritieni, A.; Fogliano, V. Characterization of a new potential functional ingredient: Coffee silverskin. *J. Agric. Food Chem.* **2004**, *52*, 1338–1343. [CrossRef]
37. Horgan, F.G.; Floyd, D.; Mundaca, E.A.; Crisol-Martínez, E. Spent Coffee Grounds Applied as a Top-Dressing or Incorporated into the Soil Can Improve Plant Growth While Reducing Slug Herbivory. *Agriculture* **2023**, *13*, 257. [CrossRef]
38. Chaklader, M.R.; Howieson, J.; Foysal, M.J.; Fotedar, R. Transformation of fish waste protein to *Hermetia illucens* protein improves the efficacy of poultry by-products in the culture of juvenile barramundi, *Lates calcarifer*. *Sci. Total Environ.* **2021**, *796*, 149045. [CrossRef]

39. Jalil, N.A.A.; Abdullah, S.H.; Ahmad, I.K.; Basri, N.E.A.; Mohamed, Z.S. Decomposition of food waste from protein and carbohydrate sources by black soldier fly larvae, *Hermetia illucens* L. *J. Environ. Biol.* **2021**, *42*, 756–761. [CrossRef]
40. Gold, M.; Ireri, D.; Zurbrügg, C.; Fowles, T.; Mathys, A. Efficient and safe substrates for black soldier fly biowaste treatment along circular economy principles. *Detritus* **2021**, *16*, 31–40. [CrossRef]
41. Broeckx, L.; Frooninckx, L.; Slegers, L.; Berrens, S.; Noyens, I.; Goossens, S.; Verheyen, G.; Wuyts, A.; Van Miert, S. Growth of black soldier fly larvae reared on organic side-streams. *Sustainability* **2021**, *13*, 12953. [CrossRef]
42. Veldkamp, T.; van Rozen, K.; Elissen, H.; van Wikselaar, P.; van der Weide, R. Bioconversion of digestate, pig manure and vegetal residue-based waste operated by black soldier fly larvae, *Hermetia illucens* L. (diptera: Stratiomyidae). *Animals* **2021**, *11*, 3082. [CrossRef]
43. Addeo, N.F.; Vozzo, S.; Secci, G.; Mastellone, V.; Piccolo, G.; Lombardi, P.; Parisi, G.; Asiry, K.A.; Attia, Y.A.; Bovera, F. Different combinations of butchery and vegetable wastes on growth performance, chemical-nutritional characteristics and oxidative status of black soldier fly growing larvae. *Animals* **2021**, *11*, 3515. [CrossRef]
44. Borel, P.; Hammaz, F.; Morand-Laffargue, L.; Creton, B.; Halimi, C.; Sabatier, D.; Desmarchelier, C. Using black soldier fly larvae reared on fruits and vegetables waste as a sustainable dietary source of provitamin a carotenoids. *Food Chem.* **2021**, *359*, 129911. [CrossRef]
45. Fischer, H.; Romano, N.; Sinha, A.K. Conversion of spent coffee and donuts by black soldier fly (*Hermetia illucens*) larvae into potential resources for animal and plant farming. *Insects* **2021**, *12*, 332. [CrossRef]
46. Villa, R.; Muñoz, H.M.B.; Jawiarczyk, N.; Vaya, A.M. Black soldier fly biorefinery: A novel upcycling route for municipal biosolids. In *Clean Energy and Resources Recovery: Biomass Waste Based Biorefineries*; Elsevier: Amsterdam, The Netherlands, 2021; Volume 1, pp. 487–500.
47. Romano, N.; Fischer, H.; Kumar, V.; Francis, S.A.; Sinha, A.K. Productivity, conversion ability, and biochemical composition of black soldier fly (*Hermetia illucens*) larvae fed with sweet potato, spent coffee or dough. *Int. J. Trop. Insect Sci.* **2022**, *42*, 183–190. [CrossRef]
48. Chemello, G.; Zarantoniello, M.; Randazzo, B.; Gioacchini, G.; Truzzi, C.; Cardinaletti, G.; Riolo, P.; Olivotto, I. Effects of black soldier fly (*Hermetia illucens*) enriched with *Schizochytrium* sp. on zebrafish (*Danio rerio*) reproductive performances. *Aquaculture* **2022**, *550*, 737853. [CrossRef]
49. Ratti, S.; Zarantoniello, M.; Chemello, G.; Giammarino, M.; Palermo, F.A.; Cocci, P.; Mosconi, G.; Tignani, M.V.; Pascon, G.; Cardinaletti, G.; et al. Spirulina-enriched Substrate to Rear Black Soldier Fly (*Hermetia illucens*) Prepupae as Alternative Aquafeed Ingredient for Rainbow Trout (*Oncorhynchus mykiss*) Diets: Possible Effects on Zootechnical Performances, Gut and Liver Health Status, and Fillet Quality. *Animals* **2023**, *13*, 173. [CrossRef]
50. Baas, J.; Schotten, M.; Plume, A.; Côté, G.; Karimi, R. Scopus as a curated, high-quality bibliometric data source for academic research in quantitative science studies. *Quant. Sci. Stud.* **2020**, *1*, 377–386. [CrossRef]
51. Mangindaan, D.; Adib, A.; Febrianta, H.; Hutabarat, D.J.C. Systematic Literature Review and Bibliometric Study of Waste Management in Indonesia in the COVID-19 Pandemic Era. *Sustainability* **2022**, *14*, 2556. [CrossRef]
52. Permana, A.D.; Ramadhani Eka Putra, J.E.N. Growth of Black Soldier Fly (*Hermetia illucens*) Larvae Fed on Spent Coffee Ground. *IOP Conf. Ser. Earth Environ. Sci.* **2018**, *187*, 012070. [CrossRef]
53. Ospina-Granobles, K.; Carrejo-Gironza, N. Efficiency of bioconversion of coffee pulp using *Hermetia illucens* (Diptera: Stratiomyidae) Larvae. *Pertanika J. Trop. Agric. Sci.* **2021**, *44*, 237–254. [CrossRef]
54. Truzzi, C.; Giorgini, E.; Annibaldi, A.; Antonucci, M.; Illuminati, S.; Scarponi, G.; Riolo, P.; Isidoro, N.; Conti, C.; Zarantoniello, M.; et al. Fatty acids profile of black soldier fly (*Hermetia illucens*): Influence of feeding substrate based on coffee-waste silverskin enriched with microalgae. *Anim. Feed Sci. Technol.* **2020**, *259*, 114309. [CrossRef]
55. Danieli, P.P.; Lussiana, C.; Gasco, L.; Amici, A.; Ronchi, B. The Effects of Diet Formulation on the Yield, Proximate Composition, and Fatty Acid Profile of the Black Soldier Fly (*Hermetia illucens* L.) Prepupae Intended for Animal Feed. *Animals* **2019**, *9*, 178. [CrossRef]
56. Milanovic, V.; Roncolini, A.; Cardinali, F.; Garofalo, C.; Aquilanti, L.; Riolo, P.; Ruschioni, S.; Corsi, L.; Isidoro, N.; Zarantoniello, M.; et al. Occurrence of antibiotic resistance genes in *Hermetia illucens* larvae fed coffee silverskin enriched with *Schizochytrium limacinum* or *Isochrysis galbana* microalgae. *Genes* **2021**, *12*, 213. [CrossRef] [PubMed]
57. Osimani, A.; Ferrocino, I.; Corvaglia, M.R.; Roncolini, A.; Milanović, V.; Garofalo, C.; Aquilanti, L.; Riolo, P.; Ruschioni, S.; Jamshidi, E.; et al. Microbial dynamics in rearing trials of *Hermetia illucens* larvae fed coffee silverskin and microalgae. *Food Res. Int.* **2021**, *140*, 110028. [CrossRef] [PubMed]
58. Pothakos, V.; De Vuyst, L.; Zhang, S.J.; De Bruyn, F.; Verce, M.; Torres, J.; Callanan, M.; Moccand, C.; Weckx, S. Temporal shotgun metagenomics of an Ecuadorian coffee fermentation process highlights the predominance of lactic acid bacteria. *Curr. Res. Biotechnol.* **2020**, *2*, 1–15. [CrossRef]
59. Hadj Saadoun, J.; Montevecchi, G.; Zanasi, L.; Bortolini, S.; Macavei, L.I.; Masino, F.; Maistrello, L.; Antonelli, A. Lipid profile and growth of black soldier flies (*Hermetia illucens*, Stratiomyidae) reared on by-products from different food chains. *J. Sci. Food Agric.* **2020**, *100*, 3648–3657. [CrossRef]
60. Martinez-Saez, N.; García, A.T.; Pérez, I.D.; Rebollo-Hernanz, M.; Mesías, M.; Morales, F.J.; Martín-Cabrejas, M.A.; del Castillo, M.D. Use of spent coffee grounds as food ingredient in bakery products. *Food Chem.* **2017**, *216*, 114–122. [CrossRef]

61. Sideris, V.; Georgiadou, M.; Papadoulis, G.; Mountzouris, K.; Tsagkarakis, A. Effect of processed beverage by-product-based diets on biological parameters, conversion efficiency and body composition of *Hermetia illucens* (L) (diptera: Stratiomyidae). *Insects* **2021**, *12*, 475. [CrossRef]
62. Khaekratoke, K.; Laksanawimol, P.; Thancharoen, A. Use of fermented spent coffee grounds as a substrate supplement for rearing black soldier fly larvae, *Hermetia illucens* (L), (Diptera: Stratiomyidae). *PeerJ* **2022**, *10*, e14340. [CrossRef]
63. Lopes, I.G.; Yong, J.W.H.; Lalander, C. Frass derived from black soldier fly larvae treatment of biodegradable wastes. A critical review and future perspectives. *Waste Manag.* **2022**, *142*, 65–76. [CrossRef] [PubMed]
64. Bohm, K.; Hatley, G.A.; Robinson, B.H.; Gutiérrez-Ginés, M.J. Black Soldier Fly-based bioconversion of biosolids creates high-value products with low heavy metal concentrations. *Resour. Conserv. Recycl.* **2022**, *180*, 106149. [CrossRef]
65. Zarantoniello, M.; Zimbelli, A.; Randazzo, B.; Compagni, M.D.; Truzzi, C.; Antonucci, M.; Riolo, P.; Loreto, N.; Osimani, A.; Milanović, V.; et al. Black Soldier Fly (*Hermetia illucens*) reared on roasted coffee by-product and *Schizochytrium sp.* as a sustainable terrestrial ingredient for aquafeeds production. *Aquaculture* **2020**, *518*, 734659. [CrossRef]
66. Milanović, V.; Cardinali, F.; Aquilanti, L.; Maoloni, A.; Garofalo, C.; Zarantoniello, M.; Olivotto, I.; Riolo, P.; Ruschioni, S.; Isidoro, N.; et al. Quantitative assessment of transferable antibiotic resistance genes in zebrafish (*Danio rerio*) fed *Hermetia illucens*-based feed. *Anim. Feed Sci. Technol.* **2021**, *277*, 114978. [CrossRef]
67. Rodrigues, D.P.; Ameixa, O.M.C.C.; Vázquez, J.A.; Calado, R. Improving the Lipid Profile of Black Soldier Fly (*Hermetia illucens*) Larvae for Marine Aquafeeds: Current State of Knowledge. *Sustainability* **2022**, *14*, 6472. [CrossRef]
68. Osimani, A.; Milanović, V.; Roncolini, A.; Riolo, P.; Ruschioni, S.; Isidoro, N.; Loreto, N.; Franciosi, E.; Tuohy, K.; Olivotto, I.; et al. *Hermetia illucens* in diets for zebrafish (*Danio rerio*): A study of bacterial diversity by using PCR-DGGE and metagenomic sequencing. *PLoS ONE* **2019**, *14*, e0225956. [CrossRef]

agriculture

MDPI

Article

Bridging the Gap: Scaling Up the Sustainable Production of the Yellow Mealworm with Agricultural By-Products—Insights into Larval Growth and Body Composition

Mariastela Vrontaki [1,*], Christina Adamaki-Sotiraki [1], Christos I. Rumbos [1], Anastasios Anastasiadis [2] and Christos G. Athanassiou [1]

[1] Laboratory of Entomology and Agricultural Zoology, Department of Agriculture, Crop Production and Rural Environment, University of Thessaly, Phytokou Str., 38446 Volos, Greece; cadamaki-s@uth.gr (C.A.-S.); crumbos@uth.gr (C.I.R.); athanassiou@uth.gr (C.G.A.)

[2] Animal Feed Anastasiadi Single Member P.C., Akropotamia, 61100 Kilkis, Greece; info@zootrofesanastasiadi.com

* Correspondence: svrontaki@uth.gr; Tel.: +30-24210-93109

Abstract: Amidst the escalating global demand for protein-rich livestock feed, there's an urgent call to explore innovative alternatives. Insects, renowned for their rich protein, lipid, and nutrient profiles, offer a sustainable solution. Integrating agricultural waste into insect diets emerges as a promising strategy to alleviate rearing costs. However, large-scale investigations into by-product valorization remain limited. Thus, our study aims to evaluate Greek agricultural by-products—brewer's spent grains, rice bran, oat and maize by-products, and animal feed mill leftovers—as potential feedstock for *Tenebrio molitor* larvae, an insect species that is authorized by the European Union for both livestock feed and human consumption. In this study, the larval growth and body composition were assessed in commercial trays, unveiling optimal performance with rice bran and brewer's spent grains. Conversely, larvae fed with animal feed mill leftovers and maize by-products displayed suboptimal outcomes. These findings underscore the potential efficacy of integrating locally produced agricultural by-products into *T. molitor* commercial production. Such an approach not only addresses the growing demand for protein-rich livestock feed but also offers a sustainable solution to agricultural waste management. In conclusion, our research contributes valuable insights towards developing economically viable insect farming.

Keywords: agricultural side-streams; by-products; insects for food and feed; larval composition; mass rearing; yellow mealworm

1. Introduction

Nowadays, the ongoing expansion of a human population urges for novel, nutritious and sustainable alternative protein sources for both human consumption and animal livestock nutrition [1–6]. Among the proposed alternative protein sources, insects emerge as a highly promising sustainable choice [4,5]. Briefly, insects have a high protein content, whereas their modest land and water requirements, coupled with their low greenhouse gas emissions and efficient feed conversion, enhance the sustainability of insect rearing compared to the traditional livestock production [4,7,8]. Along the same lines, due to their ability to biodegrade waste generated in the food industry, insect production is fully aligned with circular economy strategies that are currently promoted in the EU [9,10]. However, the complexities of their commercial production still remain a significant challenge [11].

The biggest hurdle that the insect production industry has to overcome in order to develop further is the high insect feed cost, which notably contributes to the overall production cost [12,13]. In this framework, the integration of agricultural by-products, generated during various agricultural production processes, into insect diets has been extensively examined for various edible insect species commercially produced for food and feed applications [14–19]. Extensive research has been conducted on the growth and performance of the larvae of the yellow mealworm, *Tenebrio molitor* L. (Coleoptera: Tenebrionidae) grown on a variety of agro-industrial by-products and former foodstuff products (FFPs) [14,20–24]. These included malt residual pellets, corn germ meal, sweet chestnuts, bread remains, soybeans, sweet potatoes, wheat germs, wheat, rye bran, rapeseed meal, rapeseed cake, flax, milk thistle cakes, brewery spent grains, bread and cookie leftovers, and mixes of brewer's spent grains or bread with cookies. Among these, certain substrates emerged as particularly promising based on their impact on larval growth performance, feed conversion efficiency, nutrient composition, and other parameters. Notably, malt residual pellets, corn germ meal, sweet chestnuts, bread remains, soybeans, sweet potatoes, wheat germs, and brewery spent grains demonstrated favorable outcomes in terms of larval development, biomass production, and nutrient content. These substrates offer potential as sustainable and cost-effective alternatives for the mass rearing of mealworm larvae.

Fruits and vegetables have also been utilized in *T. molitor* diets aiming to enhance insects' growth performance and nutritional value [18,25–30]. Various agri-food wastes, such as olive pomace, grape pomace, grape marcs, winery waste sludge, cucumber wastes, and tomato wastes, along with FFPs like orange peel waste and carrot pomace, have been evaluated as potential feed sources for the yellow mealworm larvae. Particularly, the inclusion of wastes like olive pomace in larval diets presented promising outcomes in terms of growth performance and nutritional properties. Similarly, the supplementation of *T. molitor* diets with vegetable and fruit wastes, such as albedo orange peel waste, carrots, and red cabbage, resulted in larvae with enhanced nutritional profiles, including increased protein content, vitamins, antioxidants, and fatty acid composition. These findings collectively underscore the feasibility and benefits of repurposing agricultural by-products and food wastes as substrates for *T. molitor* mass rearing, offering a sustainable approach to both waste management and protein production. Additionally, it is worth noting that this insect species is currently authorized within the EU not only as an ingredient of aquafeeds [31], poultry, and swine feeds [32], but also for human consumption [33].

The insect's body composition is inextricably related to the nutrients and compounds present in the consumed feed [34–37]. For instance, Ramos-Elorduy et al. [37] tested 35 dried waste organic materials as a substrate for the development of *T. molitor* larvae and suggested that the larval amino acid and nutrient content depends not only on the size and weight of the larvae, but also on the substrate's composition. Similar results were reported by Rumbos et al. [16] who evaluated the growth and body composition of the yellow mealworm larvae fed by-product-based isonitrogenous diets.

Most experiments related to *T. molitor* growth have been carried out at laboratory scale, while there are disproportionally less data about the factors that affect the performance of *T. molitor* at a large scale. Recently, Deruytter et al. [38] tested the effect of crate size, oviposition time, number of adults, and cannibalism on the reproduction of *T. molitor* in terms of the number of adults produced, oviposition time, cannibalism, and tray size at pilot scale. Similarly, two studies of Adamaki-Sotiraki et al. [39,40] conducted both at pilot scale, highlight the importance of conducting experiments at a larger scale. In this regard, the objective of the present study was to valorize at large scale Greek, locally produced, agricultural organic by-products for the rearing of *T. molitor* larvae.

2. Materials and Methods

2.1. Insect Rearing and Experimental Design

Plastic insect breeding trays (60 × 40 × 14.5 cm) (Beekenkamp Verpakkingen BV, Maasdijk, The Netherlands) were used for the stock colonies rearing, in the pilot-scale insect rearing unit of the Laboratory of Entomology and Agricultural Zoology at the University of Thessaly, in Volos, Greece, under constant conditions, i.e., continuous darkness, 60 ± 5% relative humidity and 27 ± 0.5 °C. The substrate in which larvae were reared was wheat bran, while agar cubes (20 g/L) were provided 3 times per week, as a larval moisture source.

In order to acquire newly emerged larvae, adults were allowed to oviposit for 4 d on white wheat flour. Then, the young larvae were left to feed undisturbed for 10 d on the flour. After this interval, larvae were separated from the white wheat flour by sieving. To ensure that the selected larvae had a similar age and size, only larvae that passed through a 850 μm sieve and were kept by a 600 μm sieve were used for experimentation, corresponding, according to their larval head capsule size, to 6th- and 7th-instar larvae [41]. Plastic insect breeding trays were used as experimental units and wheat bran served as control. To estimate the number of larvae, 3 subsamples, each containing at least 100 larvae, were taken and weighed, in order to estimate the weight of 10,000 larvae. Then, approximately 10,000 larvae were inserted to each tray, along with 2.1 kg of each by-product. Larvae were left to feed undisturbed for two weeks. After this interval, the content of each tray was homogenized, and a subsample was taken from each tray to determine the number of larvae per tray and the average larval weight. Attention was given, so that each subsample contained at least 100 larvae. This procedure was repeated weekly until 2 or more replicates run out of feed or more than 10% of larvae pupate. During harvesting, larvae were separated from the substrate, using 2 mm sieve. Afterward, the total amount of larvae was weighted (total harvest). The remaining substrate was also sieved, using 500 μm sieve, in order to separate the produced frass from the leftover feed, which was also weighted and recorded. Frass that was produced by T. molitor larvae passed through the 500 μm sieve, while the leftover feed was kept by the 500 μm sieve. After harvesting, larvae were fasted for 24 h and frozen (−20 °C) for further analysis. During larval development, agar was provided to larvae three times per week as a moisture source, increasing the amount of agar provided over time, always ensuring the agar availability for larvae. There were four tray replicates for each by-product treatment (oat by-product, brewer's spent grains, maize by-product, animal feed mill leftover, rice bran, and wheat bran—control).

2.2. By-Products

By-products locally produced in Greece and arising from the production processes of oats, beer, maize, rice and animal feed milling were subjected to evaluation for their suitability as *T. molitor* feed (Table 1). Specifically, the evaluation encompassed oat by-products (comprising oat straw, husks, and small oat seeds), brewer's spent grains (BSGs; in pellet form), maize by-product (resulting from the seed cleaning process), rice bran, and animal feed mill leftover (AFML), which was the residue of the animal feed grinding process (composed of 75% maize, 15% barley, 5% soft wheat, and 5% other grains). All by-products were ground using a thermomix (TM31-96 1C, Vorwerk Elektrowerke GmbH & Co. K, Wuppertal, Germany), with the exception of rice bran and the AFML, which were already in powder form. All by-products passed through a 2 mm sieve.

Table 1. Proximate composition, i.e., dry matter (%), protein (% DM), and ash (% DM) content of five agricultural by-products and wheat bran (control). For all treatments, 4 replicates were conducted (n = 4).

By-Products	Dry Matter (%)	Protein (%DM)	Ash (%DM)
Wheat bran (control)	85.8 ± 0.1^{c}	17.4 ± 0.5	4.6 ± 0.1^{d}
Oat by-product	97.1 ± 0.0^{a}	14.7 ± 2.2	6.8 ± 0.0^{c}
Brewer's spent-grains	95.6 ± 0.0^{ab}	23.4 ± 0.6	7.2 ± 0.1^{b}
Maize by-product	95.8 ± 0.1^{ab}	9.3 ± 0.6	3.4 ± 0.1^{e}
Animal feed mill leftover	93.3 ± 0.1^{bc}	5.6 ± 0.5	1.0 ± 0.0^{f}
Rice bran	91.7 ± 0.2^{bc}	17.4 ± 0.6	7.6 ± 0.1^{a}
p-value	0.006	0.11	<0.001

The case letters represent statistical differences among the larvae reared on the different by-products. In protein content (%DM), no statistically significant differences were found ($p > 0.11$). SEM stands for standard error of the mean.

2.3. Proximate Composition

Proximate composition was conducted to determine the nutrient composition of the by-products and the produced larvae. Thermal drying to constant weight in an oven at 105 °C was applied to determine the moisture content. The dried larvae were ground, in order to form an insect flour and undergo further analysis. The crude protein content was determined by Kjeldahl analysis, (Behr Labor-Technik GmbH, Düsseldorf, Germany). Four samples of larvae (0.2 g insect meal) sourced from each of the by-products were transferred to digestion test cubes. To expedite the digestion reaction, two Kjeltabs catalyst tablets (5 g Potassium Sulphate K_2SO_4 and 5 g copper (II) Sulphate $CuSO_4$, $5H_2O$) were added. Subsequently, the samples were combined with 15 mL of concentrated sulfuric acid (H_2SO_4) and placed in the Kjeltec 2000 digestion apparatus. The digestion process occurred at 150 °C for a duration of 85 min. Upon completion of the digestion process, the samples were moved to the distillation apparatus. Here, 100 mL of distilled H_2O, 80 mL of NOH, and 50 mL of H_3BO_3 were added before distillation commenced for 6 min. Finally, the titration of the resulting ammonium borate solution was conducted using a dilute hydrochloric acid solution (0.1 N). The nitrogen content of the sample (N %) was then calculated using the formula:

$$\frac{N\,\% = (mL\ HCl - mL\ blank) \times 0.8754}{Weight\ of\ the\ sample}$$

The Jones' default nitrogen-to-protein conversion factor (Kp) of 6.25 for the by-products and a specific nitrogen-to-protein conversion factor (Kp) of 4.76, for *T. molitor* larvae, were used as previously suggested [42].

The crude fat content was determined using the exhaustive Soxhlet extraction method, employing petroleum ether (40–60 °C, boiling point) in a Soxtherm Multistat/SX PC (Sox-416 Macro, Gerhard, Germany). In detail, the extraction process involved several steps. Firstly, two boiling stones were placed into extraction glass vessels, and their weights were recorded. Paper filters were then inserted into the extraction glass vessels, and 1 g of sample was weighed and deposited into the filter paper container. Next, 150 mL of petroleum ether was added to each extraction glass vessel, immersing the filter paper containers containing the sample. The extraction glass vessels, along with the filter paper, were transferred to a specialized fat extraction apparatus (Soxhlet apparatus). The extraction process proceeded in two steps: initially, the samples were heated at 150 °C in the presence of the organic solvent. Subsequently, the organic solvent was absorbed and washed into the sample for a period of 1.5 h. Afterward, the solvent was allowed to evaporate for 15 min, leaving the total lipids in the sample at the bottom of the glass extraction container. To remove any residual petroleum ether, the glass vessels were placed in an oven at 105 °C for 15 min (without the filter paper), followed by a 15 min period in a dehumidifier. Weight measurements were recorded throughout the process. The net weight of fat in the by-products was calculated using the formula:

$$Total\ lipids\ \% = \frac{Weight\ (g)\ of\ final\ extraction\ container - Weight\ (g)\ of\ original\ extraction\ container}{Weight\ (g)\ of\ the\ original\ sample} \times 100$$

Ash content was determined by dry ashing in porcelain crucibles in a muffle furnace (Nabertherm L9/12/C6, Lilienthal, Germany) at 600 °C for 5 h, and the gross energy content was determined adiabatically using an IKA oxygen bomb calorimeter (C5000; IKA Werke GmbH, Staufen, Germany) [43].

2.4. Calculations

The average individual larval weight was determined by dividing the total weight of the larvae with the number of counted larvae in each subsample. The Feed Conversion Ratio (FCR) was determined by dividing the weight of the consumed feed by the weight the larvae produced within each tray. In addition, the Efficiency of Conversion of Ingested

Food (ECI) was calculated as the weight the larvae produced within each tray divided by the consumed feed and multiplied by 100%. Finally, the specific growth rate (SGR) was calculated according to the formula SGR = 100 × (lnFBW − lnIBW)/days, where FBW and IBW stand for the final and initial body weight. FCR and SGR were calculated on a wet weight basis, whereas ECI was calculated on a dry matter basis. The Economic Conversion Ratio (ECR) was calculated by multiplying the FCR with the cost of feed per ton, and it represents the feed costs per kilogram of live larvae. At the end of the experiment, the feed with a particle size under 500 μm was considered frass. Agar was not included in the calculations.

2.5. Statistical Analysis

Statistical analysis was conducted using SPSS 26.0 (IBM Corporation, Armonk, NY, USA) for the final individual larval weight, total harvest, FCR, ECI, SGR, dry matter, lipid, ash, and protein content. All data were tested for the homogeneity and normality of variances through the Shapiro–Wilk and Levene's tests, respectively. As the by-products' ash content, ECR, and ECI data displayed a normal distribution, a parametric analysis, i.e., an analysis of variance (ANOVA), which was performed to detect the differences among treatment. The Bonferroni correction for the multiple comparisons test was performed for post hoc analysis. For the rest of the variables that did not follow a normal distribution, i.e., final individual larval weight, total harvest, FCR, SGR, larval dry matter, lipid, ash, and protein content, as well as by-products' dry matter, protein content data were subjected to non-parametric analysis. Particularly, the Kruskal–Wallis H test was employed to detect whether there were significant differences ($p < 0.05$) among the treatments, followed by Dunn multiple comparisons for post hoc testing.

3. Results

3.1. By-Products' Proximate Composition

The ash content varied significantly among the different by-products (df = (5,12), $p < 0.001$, F = 1787.6). Wheat bran (control) exhibited an ash content of 4.6%, AFML had the lowest ash content at 1.0%. Rice bran showed the highest ash content (7.6%), followed by BSG (7.2%). The oat by-product demonstrated an ash content of 6.8%, whereas the maize by-product had an ash content of 3.4%. The dry matter content differed significantly between the different by-products (df = 5, $p = 0.006$, test statistic: 16.336). Dry matter ranged between 85.8% and 97.1%, with the highest dry matter observed in the oat by-product, while the lowest was observed in what bran (control). Protein content did not present statistically significant differences ($p = 0.11$) and it was above 5.6% for all by-products.

3.2. Final Individual Larval Weight

As indicated by the individual larval weights recorded, variation in the larval growth was observed among the different dietary treatments tested (df = 5, $p < 0.001$, test statistic: 20.320). More specifically, the final individual larval weight spanned from 35.6 mg (AFML) to 98.4 mg (wheat bran) (Figure 1; Table 2). Larvae raised on wheat bran (control), BSG, and rice bran had the highest final individual larval weights. In contrast, larvae fed on the maize and the oat by-product displayed comparatively smaller growth, with the final weight remaining below 82.6 mg.

3.3. Total Harvest

The total harvest of larvae varied among the different substrates from 335.3 g to 914.4 g, with the significant differences being observed among the treatments (df = 5, $p = 0.001$, test statistic: 20.110) (Table 2). Total harvest was high for wheat bran (control) (872.3 g), BSG (914.4 g), rice bran (748.0 g), and oat by-product (708.9 g), while the lowest total harvest was recorded for AFML (335.3 g).

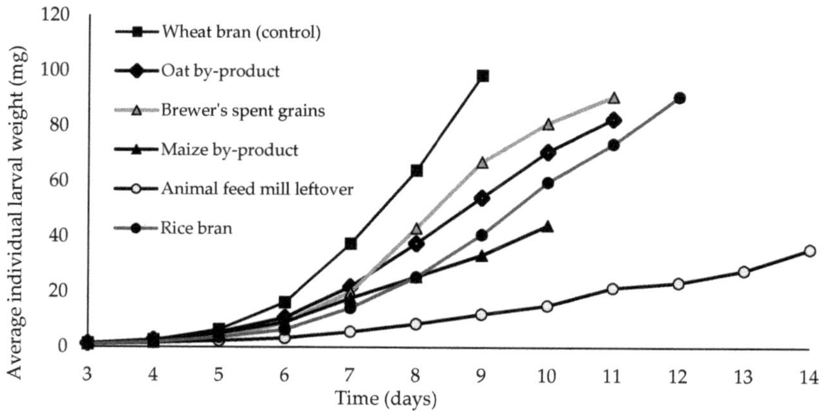

Figure 1. Average individual larval weight (mg) of *Tenebrio molitor* larvae reared on five agricultural by-products and wheat bran (control). For all treatments, 4 replicates were conducted (n = 4).

Table 2. Final individual larval weight (FILW, mg ± SEM), total harvest (g ± SEM), feed conversion ratio (FCR, ±SEM), efficiency of conversion of ingested food (ECI, % ±SEM), specific growth rate (SGR, %/day ± SEM) of *Tenebrio molitor* larvae reared on five agricultural by-products and wheat bran (control), at the end of the experiment. For all treatments, 4 replicates were conducted (n = 4).

By-Products	FILW (mg)	Total Harvest (g)	FCR	ECI (%)	SGR (%/Day)
Wheat bran (control)	98.4 ± 1.0 ^{a}	872.3 ± 33.9 ^{a}	2.3 ± 0.0 ^{abc}	43.8 ± 0.8 ^{ab}	10.7 ± 0.0 ^{a}
Oat by-product	82.6 ± 3.8 ^{bc}	708.9 ± 21.2 ^{abc}	2.0 ± 0.0 ^{a}	50.8 ± 1.1 ^{a}	8.5 ± 0.1 ^{bcd}
Brewer's spent-grains	90.7 ± 0.7 ^{ab}	914.4 ± 49.6 ^{a}	2.0 ± 0.1 ^{ab}	50.9 ± 3.3 ^{ab}	8.8 ± 0.1 ^{ab}
Maize by-product	44.1 ± 2.6 ^{bc}	434.4 ± 43.4 ^{bc}	3.9 ± 0.4 ^{cd}	26.3 ± 2.7 ^{cd}	8.6 ± 0.2 ^{abc}
Animal feed mill leftover	35.6 ± 1.9 ^{c}	335.3 ± 12.3 ^{c}	4.9 ± 0.1 ^{d}	20.4 ± 0.4 ^{d}	5.9 ± 0.0 ^{d}
Rice bran	90.7 ± 1.9 ^{ab}	748.0 ± 41.7 ^{ab}	2.8 ± 0.2 ^{bc}	36.1 ± 1.9 ^{bc}	7.8 ± 0.0 ^{cd}
p-value	0.001	0.001	<0.001	<0.001	<0.001

The case letters represent statistical differences among the larvae reared on the different by-products. SEM stands for standard error of the mean.

3.4. Feed Conversion Ratio, Efficiency of Conversion of Ingested Food, Specific Growth Rate, Economic Conversion Ratio

FCR (df = 5, $p < 0.001$, test statistic: 21.080) and ECI (df = (5, 18) $p < 0.001$, F = 41.141) varied among treatments, ranging from 2.0 to 4.9 and from 20.4 to 50.9%, respectively (Table 2). Low FCRs and high ECIs suggest the efficient utilization of the feed by the larvae. Larvae reared on BSG and oat by-product exhibited the lowest FCR and the highest ECI values (2.0 and 50.9%, 2.0 and 50.8%, respectively). Similar values were noted for larvae fed on wheat bran (control) (2.3 and 43.8%, respectively). Conversely, larvae reared on AFML resulted in high FCR (4.9) and low ECI values (20.4%), indicating the inefficient bioconversion of this by-product into body mass. SGR varied between 5.9 and 10.7%/day (df = 5, $p < 0.001$, test statistic: 21.088) (Table 2). Larvae reared on BSG (8.8%/day) and maize by-product (8.6%/day) displayed high SGR values, similar to the control (10.7%/day), while those reared on AFML, rice bran, and oat by-product exhibited lower SGR values (<8.5%/day). The economic conversion rate (ECR) of larvae varied significantly across different by-products used as feedstock (df = (5, 18), $p < 0.001$, F = 87.500) (Table 3). Larvae fed with wheat bran (control) exhibited an ECR of 342.5 EUR/ton larvae, while larvae fed with oat by-product had an ECR at 138.0 EUR/ton larvae. Treatments were grouped based on Bonferroni post hoc comparisons, with larvae reared on BSG (555.9 EUR/ton larvae) and rice bran (558.3 EUR/ton larvae) showing the highest ECR, followed by larvae reared on maize by-product (275.1 EUR/ton larvae). Larvae fed with AFML (73.6 EUR/ton larvae) demonstrated the lowest ECR among all treatments.

Table 3. Economic conversion ratio (ECR, EUR/ton larvae ± SEM) of *Tenebrio molitor* larvae reared on five agricultural by-products and wheat bran (control), at the end of the experiment and cost of by-products (price, EUR/ton). For all treatments, 4 replicates were conducted (n = 4).

By-Products	ECR (EUR/ton Larvae)	Price (EUR/ton)	Supplier
Wheat bran (control)	342.5 ± 6.5 [b]	150	Animal Feed Anastasiadi Single Member P.C.
Oat by-product	138.0 ± 6.2 [a]	70	Animal Feed Anastasiadi Single Member P.C.
Brewer's spent-grains	555.9 ± 65.0 [c]	280	Animal Feed Anastasiadi Single Member P.C.
Maize by-product	275.1 ± 58.2 [b]	70	Animal Feed Anastasiadi Single Member P.C.
Animal feed mill leftover	73.6 ± 3.2 [a]	15	Animal Feed Anastasiadi Single Member P.C.
Rice bran	558.3 ± 60.2 [c]	200	Hellenic Feedstuff Industries S.A.
p-value	<0.001		

The case letters represent statistical differences among the larvae reared on the different by-products. SEM stands for standard error of the mean.

3.5. Insects' Body Proximate Composition (Dry Matter, Lipid, Ash, and Protein Content)

The composition of larvae reared on the various by-products tested reflects significant differences in key nutritional components (dry matter content: df = 5, $p = 0.001$, test statistic: 15.700; lipid content: df = 5, $p = 0.001$, test statistic 18.748; ash content: df = 5, $p = 0.002$, test statistic: 18.676; protein content: df = 5, $p = 0.001$, test statistic: 10.057) (Table 4). Notably, larvae reared on BSG exhibited the lowest dry matter content (23.9%), similarly to the DM content of larvae fed wheat bran (33.8%). In contrast, larvae reared on AFML had the highest dry matter content (39.1%), which was not significantly different from the DM content of larvae reared on maize by-product (36.6%), rice bran (35.8%), and oat by-product (35.3%). The lipid content fluctuated between 9 and 44.3%. Larvae grown on AFML, maize, and oat by-product had a high lipid content (>39%). Conversely, BSG, wheat bran, and rice bran had the lowest lipid content (<35.7%). The ash content took its highest values for larvae raised on BSG (5.5%) and wheat bran (4.4%). Larval protein content ranged from 29.2 to 49%. Larvae raised on wheat bran and rice bran had a high protein content (37.7% and 36.7%, respectively) similarly to those raised on BSG (49%), while the lowest protein content was observed in the larvae reared on maize by-product and oat by-product (32.0% and 34.4%, respectively).

Table 4. Insects' body composition: dry matter, lipid, ash, and protein content (% DM ± SEM) of *Tenebrio molitor* larvae reared on five agricultural by-products and wheat bran (control). For all treatments, 4 replicates were conducted (n = 4), unless indicated with * (n = 3).

By-Products	Dry Matter (%)	Lipid Content (%)	Ash (%)	Protein Content (%)
Wheat bran (control)	33.8 ± 0.2 [ab]	28.2 ± 1.0 [bc*]	4.4 ± 0.2 [ab]	37.7 ± 0.9 [ab]
Oat by-product	35.3 ± 0.8 [bc]	39.0 ± 1.9 [ab*]	4.0 ± 0.1 [bc]	34.4 ± 0.4 [bcd]
Brewer's spent-grains	23.9 ± 0.4 [a]	9.0 ± 0.1 [c]	5.5 ± 0.1 [a]	49.0 ± 0.6 [a]
Maize by-product	36.6 ± 1.8 [bc]	40.5 ± 1.5 [ab*]	3.9 ± 0.1 [bc*]	32.0 ± 0.4 [cd*]
Animal feed mill leftover	39.1 ± 0.3 [c]	44.3 ± 0.5 [ab]	3.7 ± 0.0 [c]	29.2 ± 0.3 [d]
Rice bran	35.8 ± 0.1 [bc]	35.7 ± 0.3 [bc]	3.9 ± 0.1 [bc]	36.7 ± 0.2 [abc*]
p-value	0.001	0.001	0.002	0.001

The case letters represent statistical differences among the larvae reared on the different by-products. SEM stands for standard error of the mean.

4. Discussion

The upcycling of low economic value side-streams and by-products of the agri-food sector through insect bioconversion can substantially enhance the sustainable profile of the insect sector, whereas it can further contribute to the reduction in the insect production costs, improving the economic feasibility of commercial insect production. In this regard, a considerable amount of recent research works has been dedicated to the valorization of agricultural by-products as insect feedstocks [16–18,20,21,27,44]. Although of high significance, most of these studies present results generated at lab scale, which may not be directly applicable to industrial scale production systems. For instance, Yang and

Tomberlin [45] reported significant differences in the larval survival, weight, and biomass conversion of larvae of the black soldier fly, *Hermetia illucens* (L.) (Diptera: Stratiomyidae), between the benchtop and industrial treatments and suggested that the application of laboratory data to an industrial production system could lead to the underestimation of biomass production and conversion rate. Accordingly, Adamaki-Sotiraki et al. [40] evaluated the larval performance and growth of *T. molitor* grown on diets, composed of agricultural by-products, at both the laboratory and industrial scales. Although, at the laboratory scale, two diets seemed to efficiently support the larval development, only one of them resulted in a high total harvest when they were evaluated at the industrial scale. Therefore, there is a necessity for experimentation at a larger scale in order to generate data that are closer to commercial settings and could be more easily adopted by the industry. Along these lines, in the present study, we examined at large scale the potential of five agricultural by-products as single feeding substrates for *T. molitor* larvae.

Based on our results, several of the by-products tested efficiently supported the growth of *T. molitor* larvae and could be used for its rearing at commercial scale. In terms of the final individual larval weight and SGR, larvae reared on rice bran and BSG grew the best compared to larvae reared on the rest of the by-products tested and similarly to larvae reared on wheat bran, a commonly used substrate for commercial mealworm rearing. Those results are in accordance with the studies of Morales-Ramos et al. [44] and Lienhard et al. [14] that both suggested these by-products as suitable components for *T. molitor* diets. Specifically, the BSG is a by-product that has been extensively studied for its potential to be utilized as feed for *T. molitor* [23,24,26,46]. Along with the aforementioned by-products, the oat by-product gave good results regarding to the total harvest, FCR, and ECI, being in accordance with the results of previous works [16,17,44,47].

Although larvae fed the maize by-product had a high SGR, close to the SGR of larvae fed the BSG and rice bran, they had lower final individual larval weight, total harvest, and ECI, as well as a higher FCR compared to those bred in BSG and rice bran. Interestingly, our results do not coincide with the results of Lienhard et al. [14], who reported that a maize by-product, originating from the seed cleaning process, efficiently supported the development of *T. molitor* larvae, when provided as a single substrate. This may be attributed to the different experimental protocol followed in the two studies. In our study, the experiment was terminated when two or more replicates run out of feed or more than 10% pupation occurred. Regarding the maize by-product, the experiment was terminated on the grounds of feed depletion, and as a result, the larvae did not reach their optimal final weight. Among the five by-products evaluated as feedstock for *T. molitor* larvae, the worst results in terms of final individual larval weight, total harvest, FCR, ECI, and SGR were attained with the AFML. Similar results were also reported by Riudavets et al. [48] who also suggested that the *T. molitor* larvae reared on feed mill by-products performed deficiently.

The economic effectiveness of employing a specific feeding substrate is paramount, as it plays a crucial role in determining the overall financial success of insect farming. The lower ECR between the three most promising by-products of our research (rice bran, BSG, oat by-product) was emerged by the larvae reared on oat by-product (138.0 EUR/ton larvae) and it is even lower than the ECR emerged by the larvae reared on wheat bran (342.5 EUR/ton larvae)—which is widely used as *T. molitor* feed. On the other hand, the utilization of the other two promising by-products (rice bran and BSG) on which *T. molitor* larvae thrived, resulted in higher ECR (558.3 EUR/ton larvae and 555.9 EUR/ton larvae, respectively) than the ECR that occurred from wheat bran (control) utilization (342.5 EUR/ton larvae). Larvae fed with the maize by-product had a higher ECR (275.1 EUR/ton larvae) than the larvae fed with oat by-product (138.0 EUR/ton larvae), even though those substrates had the same price (70 EUR/ton). This attributes these different FCRs (FCR of larvae reared on oat by-product: 2.0, FCR of larvae reared on maize by-product: 3.9).

The chemical composition of the consumed feed is not only affecting the insects' growth, but also their body composition [22,34–37]. The highest percentage of crude protein content was found for larvae fed BSG (49.0%) and it was similar to the *T. molitor* larval protein content reported by Mancini et al. [22] and Oonincx et al. [24]. Interestingly, the BSG substrate had the highest protein content (21.3%) among the by-products tested, as well as wheat bran that was used as control (18.8%). Overall, the protein content of the larvae ranged between 29.2% and 49.0%, while it was observed that the highest the nitrogen percentage in the rearing substrate, the highest the larval protein content, as also suggested by previous studies [21,49,50]. Notably, in the case of the protein-poor AFML that only contained 5% protein, larvae fed with it had a protein content of 29.2%, which could be explained by the insects' capability to regulate their bodies' protein content, regardless of the protein content of the feed [18,36].

With respect to the larval lipid content, in most cases it was high (>28.2%), being in accordance with previous studies [16,24,50]. Conversely, larvae fed the BSG substrate had a significantly lower lipid content (9.0%), compared to the larvae reared on wheat bran (control) and the rest of the by-products. Melis et al. [23] also pointed out a noticeably lower lipid content for the larvae reared on BSG (6.39% FW), compared to the larvae reared on wheat bran (control) (12.42% FW). Finally, with regard to the dry matter content of *T. molitor* larvae, this was comparable to previously published results [16,49,50], with the relative humidity, the moisture source and feeding substrate being the major factors that affect the larval dry matter content [50].

5. Conclusions

Our investigation highlights the pragmatic feasibility of integrating oat by-product, brewery spent grains (BSG), and rice bran into the commercial production of *T. molitor* (yellow mealworm) larvae. By utilizing locally abundant agricultural residues, we not only streamline the economics of insect farming but also mitigate pressure on the agri-food sector, offering a sustainable solution for managing its by-products. This dual benefit shows the potential for synergistic relationships between the insect farming and agri-food industry, fostering a more circular and resilient approach to resource utilization. However, it is not feasible to directly translate our protocol and results to other agricultural by-products due to the inherent variability and unique characteristics of each substrate. Managing different by-products requires distinct approaches, as factors such as particle size play a crucial role. The further processing of the by-product may be necessary to ensure it is suitable for the consumption by the insects. Additionally, the efficiency of conversion from consumed feed to body mass varies depending on the specific properties of the by-product. For instance, in our experiment, we encountered challenges with the quantity of maize by-product provided to the insects, leading to a suboptimal larval weight.

In conclusion, our study emphasizes the critical importance of conducting large-scale experiments to ensure the practical applicability of research findings to industrial insect production. By bridging the gap between laboratory insights and real-world implementation, we facilitate the seamless integration of insect bioconversion technologies into commercial operations. By valorizing the agricultural by-products as valuable insect feedstocks, we not only improve the environmental footprint of food production but also create opportunities for enhanced nutritional and economic outcomes. This underscores the need for collaborative efforts between stakeholders to explore innovative solutions for sustainable resource management and pave the way for a more resilient food system.

Author Contributions: Conceptualization, C.I.R., C.G.A. and A.A.; methodology, C.A.-S. and M.V.; validation, C.G.A. and C.I.R.; formal analysis, M.V.; investigation, M.V. and C.A.-S.; resources, C.G.A. and A.A.; data curation, M.V.; writing—original draft preparation, M.V.; writing—review and editing, C.G.A., C.I.R., C.A.-S. and A.A.; visualization, C.I.R. and C.G.A.; supervision, C.I.R. and C.G.A.; project administration, C.I.R. and C.G.A.; funding acquisition, C.G.A. and C.I.R. All authors have read and agreed to the published version of the manuscript.

Funding: This research is part of the project (KMP6-0077802) entitled "EntoFEED" that is co-funded by Greece and European Union by the Action "Investment Plans of Innovation" in Central Macedonia under the framework of the Operational Program "Central Macedonia 2014–2020".

Institutional Review Board Statement: Not applicable.

Data Availability Statement: Data will be available upon request.

Conflicts of Interest: Author Anastasios Anastasiadis was employed by the company Animal Feed Anastasiadi Single Member P.C., Akropotamia Kilkis Greece. The remaining authors declare that the research was conducted in the absence of any commercial or financial relationships that could be construed as a potential conflict of interest.

References

1. Zhang, X.; Zhang, T.; Zhao, Y.; Jiang, L.; Sui, X. Structural, extraction and safety aspects of novel alternative proteins from different sources. *Food Chem.* **2023**, *436*, 137712. [CrossRef]
2. Mancini, M.C.; Antonioli, F. Italian consumers standing at the crossroads of alternative protein sources: Cultivated meat, insect-based and novel plant-based foods. *Meat Sci.* **2022**, *193*, 108942. [CrossRef] [PubMed]
3. Salter, A.M.; Lopez-Viso, C. Role of novel protein sources in sustainably meeting future global requirements. *Proc. Nutr. Soc.* **2021**, *80*, 186–194. [CrossRef] [PubMed]
4. Van Huis, A.; Gasco, L. Insects as feed for livestock production. *Science* **2023**, *379*, 138–139. [CrossRef] [PubMed]
5. FAO. *How to Feed the World in 2050, Expert Meeting on How to Feed the World in 2050*; FAO Headquarters: Rome, Italy, 2009; Available online: https://www.fao.org/fileadmin/templates/wsfs/docs/expert_paper/How_to_Feed_the_World_in_2050.pdf (accessed on 10 February 2024).
6. Kurek, M.A.; Onopiuk, A.; Pogorzelska-Nowicka, E.; Szpicer, A.; Zalewska, M.; Półtorak, A. Novel protein sources for applications in meat-alternative products—Insight and challenges. *Foods* **2022**, *11*, 957. [CrossRef]
7. Smetana, S.; Spykman, R.; Heinz, V. Environmental aspects of insect mass production. *J. Insects Food Feed* **2021**, *7*, 553–571. [CrossRef]
8. Oonincx, D.G.; De Boer, I.J. Environmental impact of the production of mealworms as a protein source for humans—A life cycle assessment. *PLoS ONE* **2012**, *7*, e51145. [CrossRef] [PubMed]
9. Van Huis, A.; Rumpold, B.A.; Van der Fels-Klerx, H.J.; Tomberlin, J.K. Advancing edible insects as food and feed in a circular economy. *J. Insects Food Feed* **2021**, *7*, 935–948. [CrossRef]
10. Frasnetti, E.; Sadeqi, H.; Lamastra, L. Integrating insects into the agri-food system of northern Italy as a circular economy strategy. *Sustain. Prod. Consum.* **2023**, *43*, 181–193. [CrossRef]
11. Bermúdez-Serrano, I.M. Challenges and opportunities for the development of an edible insect food industry in Latin America. *J. Insects Food Feed* **2020**, *6*, 537–556. [CrossRef]
12. Gasco, L.; Biancarosa, I.; Liland, N.S. From waste to feed: A review of recent knowledge on insects as producers of protein and fat for animal feeds. *Curr. Opin. Green Sustain. Chem.* **2020**, *23*, 67–79. [CrossRef]
13. Niyonsaba, H.H.; Höhler, J.; Kooistra, J.; Van der Fels-Klerx, H.J.; Meuwissen, M.P.M. Profitability of insect farms. *J. Insects Food Feed* **2021**, *7*, 923–934. [CrossRef]
14. Lienhard, A.; Rehorska, R.; Pöllinger-Zierler, B.; Mayer, C.; Grasser, M.; Berner, S. Future Proteins: Sustainable Diets for *Tenebrio molitor* Rearing Composed of Food By-Products. *Foods* **2023**, *12*, 4092. [CrossRef] [PubMed]
15. Gourgouta, M.; Rumbos, C.I.; Michail, V.; Athanassiou, C.G. Valorization of agricultural side-streams for the rearing of larvae of the lesser mealworm, *Alphitobius diaperinus* (Panzer). *Sustainability* **2022**, *14*, 7680. [CrossRef]
16. Rumbos, C.I.; Oonincx, D.G.A.B.; Karapanagiotidis, I.T.; Vrontaki, M.; Gourgouta, M.; Asimaki, A.; Mente, E.; Athanassiou, C.G. Agricultural by-products from Greece as feed for yellow mealworm larvae: Circular economy at a local level. *J. Insects Food Feed* **2022**, *8*, 9–22. [CrossRef]
17. Rumbos, C.I.; Bliamplias, D.; Gourgouta, M.; Michail, V.; Athanassiou, C.G. Rearing *Tenebrio molitor* and *Alphitobius diaperinus* larvae on seed cleaning process byproducts. *Insects* **2021**, *12*, 293. [CrossRef]
18. Ruschioni, S.; Loreto, N.; Foligni, R.; Mannozzi, C.; Raffaelli, N.; Zamporlini, F.; Pasquini, M.; Roncolini, A.; Cardinali, F.; Osimani, A. Addition of olive pomace to feeding substrate affects growth performance and nutritional value of mealworm (*Tenebrio molitor* L.) larvae. *Foods* **2020**, *9*, 317. [CrossRef]
19. Scala, A.; Cammack, J.A.; Salvia, R.; Scieuzo, C.; Franco, A.; Bufo, S.A.; Tomberlin, J.K.; Falabella, P. Rearing substrate impacts growth and macronutrient composition of *Hermetia illucens* (L.) (Diptera: Stratiomyidae) larvae produced at an industrial scale. *Sci. Rep.* **2020**, *10*, 19448. [CrossRef]
20. Bordiean, A.; Krzyżaniak, M.; Aljewicz, M.; Stolarski, M.J. Influence of different diets on growth and nutritional composition of yellow mealworm. *Foods* **2022**, *11*, 3075. [CrossRef]
21. Montalbán, A.; Sánchez, C.J.; Hernández, F.; Schiavone, A.; Madrid, J.; Martínez-Miró, S. Effects of agro-industrial byproduct-based diets on the growth performance, digestibility, nutritional and microbiota composition of mealworm (*Tenebrio molitor* L.). *Insects* **2022**, *13*, 323. [CrossRef]
22. Mancini, S.; Fratini, F.; Turchi, B.; Mattioli, S.; Dal Bosco, A.; Tuccinardi, T.; Nozic, S.; Paci, G. Former foodstuff products in *Tenebrio molitor* rearing: Effects on growth, chemical composition, microbiological load, and antioxidant status. *Animals* **2019**, *9*, 484. [CrossRef] [PubMed]

23. Melis, R.; Braca, A.; Sanna, R.; Spada, S.; Mulas, G.; Fadda, M.L.; Sassu, M.M.; Serra, G.; Anedda, R. Metabolic response of yellow mealworm larvae to two alternative rearing substrates. *Metabolomics* **2019**, *15*, 113. [CrossRef] [PubMed]
24. Oonincx, D.G.; Van Broekhoven, S.; Van Huis, A.; van Loon, J.J. Feed conversion, survival and development, and composition of four insect species on diets composed of food by-products. *PLoS ONE* **2015**, *10*, e0144601. [CrossRef] [PubMed]
25. Kotsou, K.; Chatzimitakos, T.; Athanasiadis, V.; Bozinou, E.; Adamaki-Sotiraki, C.; Rumbos, C.I.; Athanassiou, C.I.; Lalas, S.I. Waste Orange Peels as a Feed Additive for the Enhancement of the Nutritional Value of *Tenebrio molitor*. *Foods* **2023**, *12*, 783. [CrossRef] [PubMed]
26. Fasce, B.; Ródenas, L.; López, M.C.; Moya, V.J.; Pascual, J.J.; Cambra-López, M. Nutritive value of wheat bran diets supplemented with fresh carrots and wet brewers' grains in yellow mealworm. *J Insect Sci.* **2022**, *22*, 7. [CrossRef] [PubMed]
27. Brai, A.; Vagaggini, C.; Pasqualini, C.; Poggialini, F.; Tarchi, F.; Francardi, V.; Dreassi, E. Use of distillery by-products as *Tenebrio molitor* mealworm feed supplement. *J. Insects Food Feed* **2023**, *9*, 611–623. [CrossRef]
28. López-Gámez, G.; del Pino-García, R.; López-Bascón, M.A.; Verardo, V. Improving *Tenebrio molitor* Growth and Nutritional Value through Vegetable Waste Supplementation. *Foods* **2024**, *13*, 594. [CrossRef]
29. Liu, C.; Masri, J.; Perez, V.; Maya, C.; Zhao, J. Growth performance and nutrient composition of mealworms (*Tenebrio molitor*) fed on fresh plant materials-supplemented diets. *Foods* **2020**, *9*, 151. [CrossRef]
30. Rovai, D.; Ortgies, M.; Amin, S.; Kuwahara, S.; Schwartz, G.; Lesniauskas, R.; Garza, J.; Lammert, A. Utilization of carrot pomace to grow mealworm larvae (*Tenebrio Molitor*). *Sustainability* **2021**, *13*, 9341. [CrossRef]
31. European Commission (EC). EU Commission Regulation 2017/893 of 24 May 2017 Amending Annexes I and IV to Regulation (EC) No 999/2001 of the European Parliament and of the Council and Annexes X, XIV and XV to Commission Regulation (EU) No 142/2011 as Regards the Provisions on Processed Animal Protein. *Off. J. Eur. Union L* **2017**, *138*, 92–116. Available online: https://eur-lex.europa.eu/legal-content/EN/TXT/PDF/?uri=CELEX:32017R0893&from=EN (accessed on 10 February 2024).
32. European Commission (EC). Commission Regulation (EC) 2021/1372 of 17 August 2021 amending Annex IV to Regulation (EC) No 999/2001 of the European Parliament and of the Council as Regards the Prohibition to Feed Non-Ruminant Farmed Animals, Other Than Fur Animals, with Protein Derived from Animals. *Off. J. Eur. Union L* **2021**, *295*, 1–17. Available online: https://eur-lex.europa.eu/legal-content/EN/TXT/HTML/?uri=OJ:L:2021:295:FULL&from=EN (accessed on 10 February 2024).
33. European Food Safety Authority (EFSA) NDA Panel (EFSA Panel on Nutrition, Novel Foods and Food Allergens); Turck, D.; Castenmiller, J.; De Henauw, S.; Hirsch-Ernst, K.I.; Kearney, J.; Maciuk, A.; Mangelsdorf, I.; McArdle, H.J.; Naska, A.; et al. Safety of dried yellow mealworm (*Tenebrio molitor* larva) as a novel food pursuant to Regulation (EU) 2015/2283. *EFSA J.* **2021**, *19*, 6343. [CrossRef]
34. Oonincx, D.G.A.B.; Finke, M.D. Nutritional value of insects and ways to manipulate their composition. *J. Insects Food Feed* **2021**, *7*, 639–659. [CrossRef]
35. Shumo, M.; Osuga, I.M.; Khamis, F.M.; Tanga, C.M.; Fiaboe, K.K.; Subramanian, S.; Ekesi, E.; Van Huis, A.; Borgemeister, C. The nutritive value of black soldier fly larvae reared on common organic waste streams in Kenya. *Sci. Rep.* **2019**, *9*, 10110. [CrossRef] [PubMed]
36. Van Broekhoven, S.; Oonincx, D.G.; Van Huis, A.; Van Loon, J.J. Growth performance and feed conversion efficiency of three edible mealworm species (Coleoptera: Tenebrionidae) on diets composed of organic by-products. *J. Insect Physiol.* **2015**, *73*, 1–10. [CrossRef] [PubMed]
37. Ramos-Elorduy, J.; González, E.A.; Hernández, A.R.; Pino, J.M. Use of *Tenebrio molitor* (Coleoptera: Tenebrionidae) to recycle organic wastes and as feed for broiler chickens. *J. Econ. Entomol* **2002**, *95*, 214–220. [CrossRef] [PubMed]
38. Deruytter, D.; Coudron, C.L.; Teerlinck, S. Influence of crate size, oviposition time, number of adults and cannibalism on the reproduction of *Tenebrio molitor*. *J. Insects Food Feed* **2019**, *5*, 247–255. [CrossRef]
39. Adamaki-Sotiraki, C.; Deruytter, D.; Rumbos, C.I.; Athanassiou, C.G. Cross-breeding of *Tenebrio molitor* strains from a large-scale perspective. *J. Insects Food Feed* **2023**, *1*, 1–10. [CrossRef]
40. Adamaki-Sotiraki, C.; Choupi, D.; Vrontaki, M.; Rumbos, C.I.; Athanassiou, C.G. Go local: Enhancing sustainable production of *Tenebrio molitor* through valorization of locally available agricultural byproducts. *J. Environ. Manag.* **2024**, *355*, 120545. [CrossRef] [PubMed]
41. Morales-Ramos, J.A.; Kay, S.; Rojas, M.G.; Shapiro-Ilan, D.I.; Tedders, W.L. Morphometric analysis of instar variation in *Tenebrio molitor* (Coleoptera: Tenebrionidae). *Ann. Entomol.* **2015**, *108*, 146–159. [CrossRef]
42. Janssen, R.H.; Vincken, J.-P.; Van Den Broek, L.A.M.; Fogliano, V.; Lakemond, C.M.M. Nitrogen-to-protein conversion factors for three edible insects: *Tenebrio molitor*, *Alphitobius diaperinus*, and *Hermetia illucens*. *J. Agric. Food Chem.* **2017**, *65*, 2275–2278. [CrossRef] [PubMed]
43. Karapanagiotidis, I.T.; Psofakis, P.; Mente, E.; Malandrakis, E.; Golomazou, E. Effect of fishmeal replacement by poultry by-product meal on growth performance, proximate composition, digestive enzyme activity, haematological parameters and gene expression of gilthead seabream (*Sparus aurata*). *Aquac. Nutr.* **2018**, *25*, 3–14. [CrossRef]
44. Morales-Ramos, J.A.; Rojas, M.G.; Kelstrup, H.C.; Emery, V. Self-selection of agricultural by-products and food ingredients by *Tenebrio molitor* (Coleoptera: Tenebrionidae) and impact on food utilization and nutrient intake. *Insects* **2020**, *11*, 827. [CrossRef] [PubMed]
45. Yang, F.; Tomberlin, J.K. Comparing selected life-history traits of black soldier fly (Diptera: Stratiomyidae) larvae produced in industrial and bench-top-sized containers. *J. Insect Sci.* **2020**, *20*, 25. [CrossRef]

46. Kim, S.Y.; Kim, H.G.; Lee, K.Y.; Yoon, H.J.; Kim, N.J. Effects of Brewer's spent grain (BSG) on larval growth of mealworms, *Tenebrio molitor* (Coleoptera: Tenebrionidae). *Int. J. Indust Entomol. IJIE* **2016**, *32*, 41–48. [CrossRef]
47. Kröncke, N.; Benning, R. Self-selection of feeding substrates by *Tenebrio molitor* larvae of different ages to determine optimal macronutrient intake and the influence on larval growth and protein content. *Insects* **2022**, *13*, 657. [CrossRef]
48. Riudavets, J.; Castañé, C.; Agustí, N.; Del Arco, L.; Diaz, I.; Castellari, M. Development and biomass composition of *Ephestia kuehniella* (Lepidoptera: Pyralidae), *Tenebrio molitor* (Coleoptera: Tenebrionidae), and *Hermetia illucens* (Diptera: Stratiomyidae) reared on different byproducts of the agri-food industry. *J. Insect Sci.* **2020**, *20*, 17. [CrossRef]
49. Rumbos, C.I.; Karapanagiotidis, I.T.; Mente, E.; Psofakis, P.; Athanassiou, C.G. Evaluation of various commodities for the development of the yellow mealworm, *Tenebrio molitor*. *Sci. Rep.* **2020**, *10*, 11224. [CrossRef] [PubMed]
50. Kröncke, N.; Benning, R. Determination of moisture and protein content in living mealworm larvae (*Tenebrio molitor* L.) using near-infrared reflectance spectroscopy (NIRS). *Insects* **2022**, *13*, 560. [CrossRef] [PubMed]

agriculture

MDPI

Article

Lactic Acid Bacteria from *Bombyx mori* Frass: Probiotic Properties and Antagonistic Activities

Siripuk Suraporn [1], Sopa Cansee [2], Sebastian Hupfauf [3] and Thomas Klammsteiner [4,*]

[1] Department of Biology, Faculty of Science, Mahasarakham University, Kantarawichai District, Maha Sarakham 41150, Thailand; siripuk.s@msu.ac.th
[2] Faculty of Engineering, Mahasarakham University, Kantarawichai District, Maha Sarakham 41150, Thailand; sopa.c@msu.ac.th
[3] Department of Microbiology, University of Innsbruck, 6020 Innsbruck, Austria; sebastian.hupfauf@uibk.ac.at
[4] Department of Ecology, University of Innsbruck, 6020 Innsbruck, Austria
* Correspondence: thomas.klammsteiner@uibk.ac.at

Abstract: Insect farming is gaining attention as a promising area for exploring probiotic bacteria, which can benefit both insect health and various industries. Silkworm farming is a key industry in Thailand; however, challenges such as disease susceptibility and optimising growth require innovative solutions for sustainable practices. Our study addresses this by assessing lactic acid bacteria (LAB) in native Thai silkworm faeces, which accumulate as natural by-products during the rearing process. We conducted biochemical tests, including those for catalase, haemolytic activity, bile salt tolerance, antimicrobial activity, antibiotic susceptibility, and cell surface hydrophobicity, along with taxonomic classification. Out of 102 isolates, eight potential probiotics were selected, with five showing strong probiotic traits like acid and bile salt tolerance and cell surface hydrophobicity, enhancing gut survivability. These isolates also displayed antagonistic activity against pathogens like *Staphylococcus aureus*, *Salmonella typhimurium*, *Escherichia coli*, and *Pseudomonas aeruginosa*. Safety assessments confirmed their safety, with no haemolytic activity and sensitivity to antibiotics like chloramphenicol and amoxicillin. These LAB isolates (SP04, SP06, SP44, SP64, and SP67), identified as *Enterococcus faecalis* strain NBRC 100481, show promise as in vitro probiotics for silkworm rearing, calling for further in vivo evaluation.

Keywords: rearing optimisation; alternative protein source; insect biotechnology; artificial diet; entomophagy; lactic acid bacteria; lactobacillus characteristics; probiotics; sericulture; microbial ecology

1. Introduction

As the global population continues to increase, the popularity of insects for both food and feed applications is steadily increasing. This trend is gaining momentum because of their remarkable nutritional profile and sustainable production practices, making insects a compelling solution to the challenges posed by the growing world population [1]. Insect farming may not only support protein self-sufficiency but can also offer novel means for organic waste management and the generation of value-added products, including fertilisers from rearing residues [2]. *Bombyx mori* (Lepidoptera: Bombycidae; Linnaeus, 1758) is one of the most important farmed insects in Thailand [3]. Throughout its life cycle, the silkworm undergoes various stages, starting as an egg, progressing through five larval instars, transforming into a pupa, and finally emerging as an adult silk moth (Figure 1). The polyvoltine silkworm can be reared year-round and is primarily found in tropical regions, such as Thailand. Beyond its traditional use in silk production, *B. mori* serves as a valuable insect model in life sciences and plays an important role in antiviral agent screening (e.g., silkworm–baculovirus infection model) [4] and in environmental monitoring [5], and also as food [6] and feed [7].

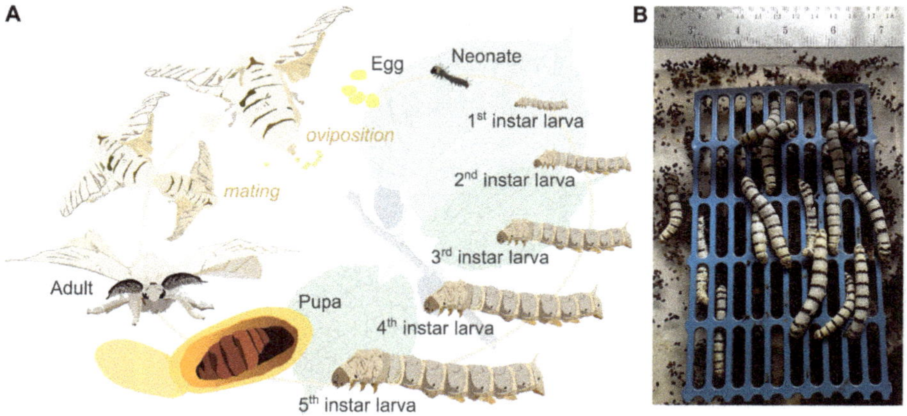

Figure 1. (**A**) Silkworm (*Bombyx mori*) life cycle and (**B**) faeces harvested under aseptic conditions in a laminar flow cabinet.

Only the monophagous larvae of *B. mori* exclusively feed on mulberry (*Morus* spp.) leaves, generating faeces that are rich in nutrients and microorganisms [8]. This waste material represents a significant by-product of silkworm rearing and is produced in large quantities in Southeast Asian countries, including Thailand. Silkworm faeces, rich in (plant) nutrients and mixed with undigested substrate residues (frass), are used as organic fertilisers and in biogas production [9–11]. In Thailand, silkworm frass serves as a vital fertiliser for mulberry plantations, offering organic nutrients and showcasing sustainable agricultural practices. China has leveraged silkworm faeces in traditional medicine, emphasising their high fat, protein, and amino acid contents [12]. Reports indicate its use in pharmaceutical supplements and the food industry [13], such as oral iron supplementation, to ameliorate iron deficiency associated with anaemia [14]. Moreover, silkworm faeces have been shown to have antiviral and anti-inflammatory effects and can be used to treat symptoms related to rheumatoid arthritis [6]. Silkworm faeces primarily consist of organic matter and ash, constituting about 84–90% and 16–20% of their biomass, respectively [15]. In particular, faeces from silkworms reared on a natural mulberry leaf diet show an enriched content of amino acids, carbohydrates, and lipids [16].

Insects are generally considered a rich source of probiotic microorganisms that reside in their guts, where they provide metabolic support and defence mechanisms that support insect health [17]. These microbes are promising targets for biotechnological applications and have previously been isolated based on their strong cellulolytic [18], amylolytic [18], and xylolytic [19] properties. A metagenomic study by Yeruva et al. [20] revealed diverse patterns of microbial communities across various silkworm breeds, highlighting the need for further in-depth characterisation of silkworm-derived probiotics.

This study focused on the potential of lactic acid bacteria (LAB) isolated from faeces of a native Thai silkworm strain, considering the beneficial probiotic qualities of many LABs. Dong et al. [21] revealed that *B. mori* reared on fresh mulberry leaves had a higher bacterial diversity in their guts than those reared on an artificial diet. LAB, particularly those from the *Lactobacillus* genus, are widely acknowledged for their positive impact on host health, interacting with the gastrointestinal mucosa, stimulating the immune system, and decreasing pathogen infections [22]. Bermudez-Brito et al. [23] demonstrated and summarised six major mechanisms of action of probiotics: (a) enhancement of the epithelial barrier, (b) increased adhesion to the intestinal mucosa, (c) inhibition of pathogen adhesion, (d) competitive exclusion of pathogenic microorganisms, (e) production of antimicrobial substances, and (6) modulation of the immune system. Therefore, defining probiotic properties involves multiple screening criteria, such as tolerance to low pH and bile salts, inhibition of pathogens, cell surface hydrophobicity, safety assessments based on haemolytic activity, susceptibility to antibiotics, and the potential to induce host immune responses [24,25]. Bacteria belonging to the enterococci genus are widely recognised for their probiotic benefits and play a crucial role as food additives for human consumption [26]. Specifically, *E. faecalis* isolated from the meconium of newborns possesses notable probiotic properties suitable for applications that foster human health [27]. Enterococci have been extensively used to promote animal health and husbandry practices [28]. As feed additives, enterococci contribute to the induction of the immune system, thereby enhancing resistance against pathogens and supporting overall health.

Given the medicinal properties of silkworm faeces, this study aimed to isolate and screen LAB from the larval faeces of a Thai *B. mori* strain, with a focus on characterising their probiotic potential.

2. Materials and Methods

2.1. Silkworm Rearing and Harvesting of Fresh Faeces

Silkworm eggs of a Thai strain were purchased from a silkworm farm in the Kudrung District, Mahasarakham Province, Thailand. The eggs were incubated at 25–26 °C for 10 days. After hatching, the larvae were fed with mulberry leaves thrice a day (at 06:00, 12:00, and 18:00). Faecal samples were collected from 100 larvae on day 4 of the 5th instar (Figure 1A) under sterile conditions in a laminar flow cabinet to avoid cross-contamination (Figure 1B).

2.2. Isolation of Lactic Acid Bacteria (LAB) and Assessment of Acid Tolerance

The inoculum was prepared by mixing 10 g of silkworm faeces with 90 mL of sterile 0.9% saline solution. De Man–Rogosa–Sharpe (MRS) broth (pH 2.5) was inoculated with 100 µL of the silkworm faeces suspension and incubated at 37 °C for 2 h to determine the acid tolerance of lactic acid bacteria [29]. After the acid treatment, the samples were serially diluted from 10^{-1} to 10^{-4}. Each dilution was plated in four replicates by spreading 100 µL on MRS agar (pH 6.2) supplemented with 1% $CaCO_3$ and 0.004% bromocresol purple (BCP). The plates were incubated at 37 °C for 48 h under anaerobic conditions using an Anaeropack® system (Mitsubishi Gas Chemical Co. Inc., Tokyo, Japan). LAB colonies exhibiting characteristic morphological features and producing a yellow colour from BCP were selected for the screening of probiotic properties. Of the 102 colonies, 74 changed their colour to yellow and were surrounded by a clear zone.

2.3. Gram Staining and Catalase Test

Gram staining was conducted by transferring a bacterial culture onto a glass microscope slide using a sterilised inoculation loop and smearing it with water. The culture was stained with crystal violet for 1 min, followed by fixing with iodine dye for an additional minute. Discolouration was achieved by washing with 70% ethanol. Finally, the culture was counterstained with safranin for 30 s and then rinsed with water. Morphological features of the bacteria were observed under a light microscope at 40× magnification.

For the catalase test, the bacterial culture was smeared onto a dry and cleaned slide using a sterilised inoculating loop, and a drop of hydrogen peroxide was added to the culture. Immediate observation of bubbles indicated a positive catalase reaction, whereas the absence of bubbles indicated a negative catalase reaction. Only colonies showing gram-positive and catalase-negative reactions were selected. The selected colonies were then evaluated for probiotic properties and stored at −20 °C in MRS broth containing 20% glycerol for further studies.

2.4. Haemolytic Activity

The haemolytic activity of LAB was assessed following a modified protocol using human blood [30,31]. Overnight cultures of each LAB isolate in MRS broth were streaked onto Columbia agar plates (BD Difco, Thermo Fisher Scientific, Waltham, MA, USA) supplemented with 5% human blood. The plates were incubated at 37 °C for 48 h. Subsequently, the bacterial plates were examined for the presence of haemolytic zones, categorised as alpha haemolysis (greenish zone), beta haemolysis (clear zone), or gamma haemolysis (no haemolysis) around the LAB colonies. Only the strains exhibiting gamma haemolysis were considered safe. The selected LAB isolates were then subjected to antibiotic assays, with *Staphylococcus aureus* serving as a positive control, showing a visible clear zone for beta haemolysis [32].

2.5. Bile Salt Tolerance Assay

The survival of LAB under acidic conditions with bile salts was also investigated. The bile salt tolerance of the isolates was evaluated by adding 0.3% (w/v) bile salt (Oxgall, Himedia, Maharashtra, India) to MRS broth (pH 2.5). For the test, a 5 mL culture of LAB isolates grown overnight in MRS broth at 37 °C under anaerobic conditions was used. Viable LAB colonies were determined using serial dilution and plate counting. MRS broth without LAB was used as a negative control. Survival rate was assessed by contrasting counts of viable colonies (mean log(cfu)/mL) before and after exposure to bile salts using the following Equation (1):

$$Survival\ rate\ (\%) = \frac{\log\ value\ of\ cells\ survived}{\log value\ of\ initially\ viable\ cells} \times 100 \quad (1)$$

2.6. Cell Surface Hydrophobicity Assay

To assess the bacterial adhesion of LAB isolates to solvents, cell surface hydrophobicity was determined following established protocols [33,34]. LAB suspensions were prepared in phosphate-buffered saline (PBS) and centrifuged at 6500 rpm for 5 min. The PBS-resuspended LAB were mixed with an equal volume of physiological saline to obtain an optical density of 0.5 at a wavelength of 600 nm (OD600). Three mL of this suspension was transferred to an autoclaved tube and 1 mL of xylene was added. After shaking on a vortex shaker for 1 min, the tubes were left undisturbed for 15 min to facilitate phase separation. Subsequently, the absorbance of the aqueous phase was measured at a wavelength of 600 nm. Hydrophobicity was calculated from three replications, representing the percent decrease in the OD of the original bacterial suspension due to LAB partitioning into the hydrocarbon layer [35]. The hydrophobicity of the cells was measured at OD600, and the cell surface hydrophobicity ratio (%) was calculated using the following Equation (2) [29]:

$$\text{Cell surface hydrophobicity}(\%) = \frac{OD600\ before\ mixing - OD600\ after\ mixing}{OD600\ before\ mixing} \times 100 \quad (2)$$

2.7. Evaluation of LAB Isolates against Bacteria Pathogens

The antibacterial activities of the isolates against four human pathogens (*Escherichia coli* ATCC 25922, *Salmonella typhimurium* ATCC 14028, *Pseudomonas aeruginosa*, and *Staphylococcus aureus* ATCC 25923) were assessed using the agar spot method [36]. Sixteen LAB isolates were spotted onto the surface of MRS agar and incubated at 37 °C for 24 h under anaerobic conditions. After culturing, 50 µL of each bacterial pathogen was inoculated into a Brain Heart Infusion (BHI, 0.5% soft agar) and poured onto the MRS agar plates. The plates were incubated at 37 °C for another 24 h, and inhibition activity was observed. The diameter of the clear zone for each isolate against each indicator pathogen was measured, and inhibition levels were classified as (−) for no inhibition, (+) for 0.5–6 mm, (++) for 7–12 mm, and (+++) for more than 12 mm inhibition [37].

2.8. Sensitivity of LAB to Antibiotics

Antibiotic susceptibility was assessed using the Bauer–Kirby disk diffusion method [38] with commercial antibiotic discs (Himedia, Maharashtra, India). Five antibiotics, penicillin (10 units), chloramphenicol (30 µg), gentamicin (10 µg), tetracycline (30 µg), and amoxicillin (30 µg), were tested [39,40]. LAB isolates were swabbed on Mueller–Hinton agar, and antibiotic discs were placed on the plate before incubation at 37 °C for 24 h. The inhibition zone around each antibiotic disc was measured, and LAB sensitivity was categorised as sensitive ("S"), intermediate ("I"), or resistant ("R") based on the enclosed zone size interpretative charts (Himedia): amoxicillin (S $\geq$ 18 mm; I = 14–17 mm; R $\leq$ 13 mm), chloramphenicol (S $\geq$ 18 mm; I = 13–17 mm; R $\leq$ 12 mm), gentamicin (S $\geq$ 15 mm; I = 13–14 mm; R $\leq$ 12 mm), penicillin (S $\geq$ 15 mm; R $\leq$ 14 mm), and tetracycline (S $\geq$ 15 mm; I = 12–14 mm; R $\leq$ 11 mm).

2.9. Colony PCR, Sanger Sequencing, and Taxonomic Identification

For a single colony PCR, a mixture of 0.5 µL 27F forward primer (5′-AGAGTTTGATCA TGGCTCA-3′), 0.5 µL 1492R reverse primer (5′-TACGGTTACCTTGTTACGACTT-3′), 0.5 µL 2% BSA, and 12.5 µL RedTaq Polymerase 2× master mix (containing 1.5 mM $MgCl_2$, VWR, Radnor, PA, USA) was prepared and combined with 11 µL PCR-grade H_2O. Bacterial isolates selected from the enrichment plates were carefully transferred into reaction tubes containing 25 µL of the PCR mix using sterile pipette tips. Colony PCR was performed according to the thermocycler program described in Table S1. Subsequently, the PCR products were purified using the GeneElute PCR Clean-up Kit (Sigma-Aldrich, St. Louise, MO, USA). Prior to Sanger sequencing of the entire 16S rRNA gene, 2 µL of the 27F forward primer was added to each purified PCR product. The resulting full-length 16S reads were subjected to nucleotide BLAST search for taxonomic assignment. All the samples were amplified and sequenced in duplicate for more reliable verification.

2.10. Statistical Analysis

The experiments were carried out in triplicate (n = 3), and the results are presented as the mean ± standard deviation. Data analysis was performed in R (v. 4.3.1) using one-way analysis of variance, and statistical significance was defined as $p < 0.05$. Prior to analysis, normality of the data was checked using the Shapiro–Wilk test. Tukey's Honest Significant Difference post hoc test was used for pairwise comparisons of groups.

3. Results

3.1. Morphological and Physiological Study of Lactic Acid Bacteria

Through morphological analysis of the colonies, 102 LAB colonies derived from the faeces of a local Thai silkworm strain were identified. Initial characterisation of these isolated LAB involved criteria such as gram positivity, exhibiting either cocci or rod shapes (Figure 2A,B) and demonstrating survival in an acidic environment (pH 2.5) through the hydrolysis of $CaCO_3$ combined with bromocresol purple on MRS agar. Notably, 74 isolates exhibited morphological features indicative of their potential as probiotic candidates, specifically for gram-positive bacteria. Additionally, all isolated strains demonstrated defence mechanisms against oxidative stress, as evidenced by a catalase-positive reaction.

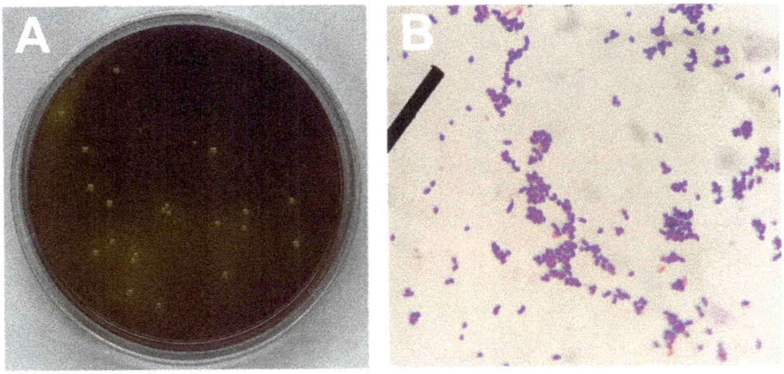

Figure 2. Preliminary characterisation of lactic acid bacteria (LAB) isolated from faeces of silkworms: (**A**) single colonies of LAB isolated on selective MRS agar, (**B**) microscopic (40× magnification) observation of gram-positive LAB.

3.2. Haemolytic Activity, Bile Salt Tolerance, and Cell Surface Hydrophobicity

To assess the safety of the selected LAB, the haemolytic reaction was determined by streaking LAB on Columbia blood agar containing 5% (*v*/*v*) human blood. Out of the 74 isolates, 49 isolates did not exhibit a halo zone, confirming that they were gamma haemolytic. In contrast, a clear haemolytic zone (beta haemolysis) was observed around *S. aureus*, which served as positive control.

Among the LAB isolates, SP04, SP06, SP44, SP64, and SP67 displayed notable acid and bile salt tolerance (Figure 3A). After 24 h in bile salt conditions, their survival rates were 26.3 ± 0.6%, 51.2 ± 1.0%, 12.8 ± 0.3%, 72.9 ± 0.8%, and 33.3 ± 0.0%, resulting in highly significant differences across the isolates ($F(4,10) = 3938$, $p < 0.001$ ***).

Our study revealed that LAB isolates exhibited hydrophobicity ranging from 41% to 65% on average (Figure 3B). Notably, we observed significant variation among replicates within each isolate yet detected no significant differences across distinct isolates ($F(4,10) = 0.141$, $p = 0.94$).

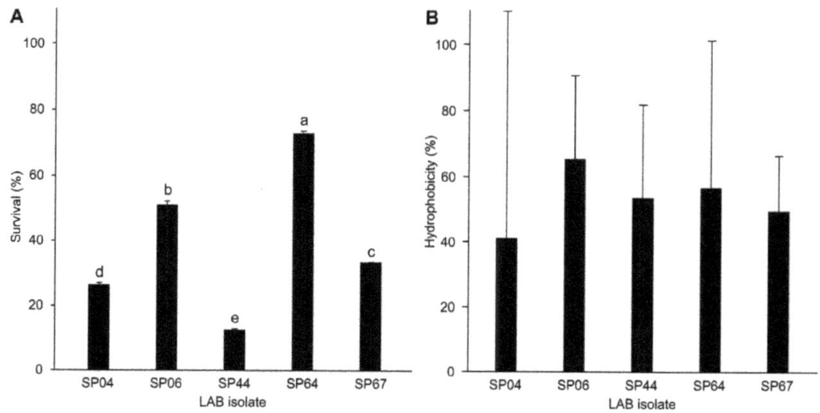

Figure 3. (**A**) Selected lactic acid bacteria (LAB) isolates' tolerance to bile salt (n = 3) (ANOVA: $F(4,10) = 3938$, $p < 0.001$). (**B**) Cell surface hydrophobicity of selected LAB (n = 3). No significant differences between LAB were observed in terms of hydrophobicity (ANOVA: $F(4,10) = 0.141$, $p = 0.94$). Values are presented as mean ± standard deviation. Lower-case letters indicate statistically significant differences between groups.

3.3. Selected LAB Isolates against Bacterial Pathogens

The selected LAB isolates were subjected to an agar spot test on MRS agar against indicator bacteria, demonstrating their inhibitory effects on pathogenic strains. The degree of inhibition, measured by the clear zone, varied between 4.2 mm and 19.8 mm, as detailed in Table 1.

Table 1. Antimicrobial testing of selected lactic acid bacteria (LAB) against pathogenic bacteria in MRS broth (pH 6.5) (n = 3). Evaluation based on Vélez et al. (2007) [37]: 0.5–6 mm = (+), 7–12 mm = (++), >12 mm = (+++).

LAB Isolate	Pathogenic Bacteria [ø mm]			
	Escherichia coli	*Staphylococcus aureus*	*Salmonella typhimurium*	*Pseudomonas aeruginosa*
SP04	5.3 ± 0.5 (+)	12.8 ± 1.0 (+++)	19.8 ± 0.8 (+++)	14.5 ± 0.0 (+++)
SP06	4.8 ± 0.3 (+)	12.8 ± 0.8 (+++)	18.5 ± 1.4 (+++)	14.2 ± 0.3 (+++)
SP44	13.1 ± 0.3 (+++)	4.2 ± 0.5 (+)	14.2 ± 0.5 (+++)	6.6 ± 0.4 (++)
SP64	12.9 ± 0.5 (+++)	7.3 ± 0.4 (++)	19.7 ± 1.1 (+++)	11.3 ± 0.4 (++)
SP67	17.9 ± 0.9 (+++)	9.0 ± 0.3 (++)	16.3 ± 0.3 (+++)	13.3 ± 0.7 (+++)

3.4. Antibiotic Susceptibility Test

To confirm the suitability of probiotic candidates for use as feed supplements, it is crucial to evaluate the sensitivity of the isolates to a relevant range of antibiotics of both human and animal importance. The five selected LAB isolates were subjected to antibiotic susceptibility testing using various antibiotics. In line with the guidelines by the European Food Safety Authority, two antibiotic groups were considered: (I) agents inhibiting cell wall synthesis (amoxicillin and penicillin) and (II) agents inhibiting protein synthesis (chloramphenicol, gentamicin, and tetracycline). Notably, the tested isolates demonstrated sensitivity to amoxicillin, chloramphenicol, penicillin, and tetracycline as denoted by the letter 'S' (sensitive). Four out of five isolates, however, showed resistance ('R') against gentamicin, and for isolate SP44, this antibiotic showed intermediate inhibitory ('I') effects, as detailed in Table 2.

Table 2. Antibiotic susceptibility of selected lactic acid bacteria (LAB) isolates. Data on the measured inhibition zones are expressed as mean diameter ± standard deviation (n = 3). Susceptibility was determined based on the size interpretative charts enclosed by the antibiotic disc manufacturer (Himedia, Maharashtra, India). (R) = resistant, (S) = sensitive, (I) = intermediate inhibition.

LAB Isolate	Antibiotic Susceptibility Zone of Inhibition (mm)				
	Amoxicillin	Chloramphenicol	Gentamicin	Penicillin	Tetracycline
SP04	27.3 ± 1.9 (S)	20.7 ± 1.0 (S)	12.2 ± 0.3 (R)	19.5 ± 1.3 (S)	23.3 ± 0.8 (S)
SP06	29.2 ± 0.3 (S)	20.3 ± 0.8 (S)	12.3 ± 0.6 (R)	19.8 ± 0.6 (S)	23.8 ± 2.4 (S)
SP44	28.0 ± 0.5 (S)	20.0 ± 1.0 (S)	13.3 ± 0.6 (I)	19.3 ± 0.6 (S)	23.3 ± 0.8 (S)
SP64	27.7 ± 1.5 (S)	19.5 ± 0.9 (S)	12.8 ± 1.4 (R)	18.8 ± 0.3 (S)	22.7 ± 0.8 (S)
SP67	27.8 ± 0.8 (S)	20.0 ± 0.0 (S)	12.3 ± 0.6 (R)	20.3 ± 1.0 (S)	19.3 ± 4.6 (S)

3.5. Genetic Identification

The sequences derived from Sanger sequencing of the five isolates were all identified as *Enterococcus faecalis* strain NBRC 100481 (Accession no.: NR_113902.1). Comprehensive information, including raw sequences and detailed nucleotide BLAST output, is provided in Table S2.

4. Discussion

In the context of sericulture, silk production has been promoted and developed to support a circular economy. Recently, the silk-producing industry has found applications beyond traditional clothing, extending to uses in human food and animal feed [41]. As a by-product of insect mass-rearing, insect faeces have gained popularity, particularly as a novel agricultural fertiliser, due to their vast microbial and nutritional diversity and soil-enhancing properties [42,43]. In this study, we aimed to harness the potential of waste from silkworm farming by isolating LAB from the faeces of a Thai strain of *B. mori* and investigate its potential use as probiotics. The results indicated that these LAB isolates met the probiotic criteria established by Bermudez-Brito et al. (2012) [23]. Preliminary isolation tests were conducted using MRS nutrient media (MRS agar and MRS broth) at an acidified pH of 2.5. These tolerant isolates were further assessed for their probiotic properties. Five selected isolates demonstrated bile salt hydrolase activity and the ability to adhere to the epithelial cells. They were sensitive to amoxicillin, chloramphenicol, penicillin, and tetracycline, but showed resistance to gentamicin, indicating that their growth could be inhibited by a large selection of antibiotics if necessary. Based on their properties, the isolated LAB can be considered potential candidates for probiotics in animal feed [44]. However, caution should be paid to resistance to, e.g., aminoglycoside antibiotics such as gentamicin, which is frequently observed in LAB isolated from farmed animals [45]. The exposure of these animals to antibiotics increases the selective pressure on microbial communities in the host guts, favouring the proliferation of resistant microbes and driving the acquisition of resistances via horizontal gene transfer [45]. Artificial environments such as insect mass-rearing facilities may have similar effects on microbial populations. Although LAB, such as enterococci isolated from wild animals, also demonstrate resistances to antibiotics; farmed animals, pets, and humans constitute their primary reservoir [46]. Therefore, special attention should be directed towards monitoring the spread of resistances and associated virulence factors.

Sanger sequencing identified the isolates as *Enterococcus faecalis* strain NBRC 100481, a microaerophile mesophilic representative of the family Enterococcaceae. Recently, *E. faecalis* was found to be an effective antagonist against infections in *B. mori* caused by Microsporidia, including *Nosema bombycis*, the pathogen responsible for Pébrine disease in their larvae [47,48]. Enterococci are also capable of causing opportunistic infections [49] but are typically known for their probiotic effects when present as commensal gut colonisers. In healthy silkworms, enterococci make up a considerable amount of the gut microbiota, indicating their relevance to host health [50]. However, other representatives of enterococci such as *Enterococcus mundtii*, can have adverse effects on *B. mori* larvae and cause the flacherie disease [51], while the same species has been proven to protect another insect species, *Tribolium castaneum*, against *Bacillus thuringiensis* [52].

A critical aspect of probiotic bacteria is their ability to survive in acidic conditions in the gut. In our investigation involving 102 LAB isolates, all met the primary criteria for probiotic properties, displaying gram-positivity and growth in MRS broth with an acidic pH of 2.5 to a neutral pH of 6.7. These findings align with previous reports emphasising the importance of LAB viability in low-pH environments [53–55].

In this study, 102 different isolates were identified, of which 74 were found to be gram-positive and tested negative for catalase, as indicated by the absence of bubbles [56]. Out of these 74 isolates, 49 were cocci-shaped and exhibited gamma haemolysis. Additionally, bacterial cell surface hydrophobicity, assessed through auto-aggregation, demonstrated strong hydrophobicity to the hydrocarbon xylene, further supporting their potential as indigenous probiotics in the host gut [57,58].

Considering the importance of antibiotic resistance in probiotics, our results showed that the five LAB isolates identified as probiotic candidates exhibited susceptibility to common antibiotics [6]. This suggests that probiotics isolated from the faeces of silkworm larvae are safe for use in silkworm farming, benefiting from their origin in the native host's gut environment and potentially offering advantages over probiotics isolated from other sources [59]. Moreover, these five isolates exhibited strong inhibitory effects on common pathogens such as *E. coli*, *S. aureus*, *S. typhimurium*, and *P. aeruginosa*. Enterococci are ubiquitous and can act as both probiotics or opportunistic pathogens depending on the strain and environmental conditions [60], though the specific mechanisms of their pathogen inhibition were not covered in this study. However, previous research has shown that *E. faecalis* can effectively colonise the intestine and exhibits immunoregulatory activities, including the activation of immune system cells [60]. Effective probiotics typically adhere to the gut lining, providing a barrier and excreting antimicrobial peptides against pathogens [61].

In this study, LAB were isolated from the faeces of a polyvoltine Thai silkworm strain. Previous research has shown a higher prevalence of *Enterococcus* sp. in polyvoltine than in bivoltine silkworms [20]. Probiotic criteria include the ability to reduce midgut pH from alkaline to acidic, which is a key factor in preventing infectious diseases caused by microorganisms, such as viruses, bacteria, and microsporidia.

Lately, lactic acid bacteria have been increasingly used as supplements and food additives in animals, including insects [19,62]. However, the mechanisms and interactions between silkworms and LAB remain unclear, and further research of LAB survival in the gut is necessary. Our results suggest that LAB supplementation may offer opportunities to enhance health while potentially reducing pathogens, particularly in preventing viral and bacterial infections in silkworms, an area with limited reports to date.

5. Conclusions

Our study was designed to capitalise on the utilisation of waste from silkworm farming. Lactic acid bacteria isolates were specifically selected from the faeces of a local strain of *B. mori* in Thailand for their promising probiotic potential. The five selected isolates demonstrated bile salt hydrolase activity and the ability to adhere to epithelial cells. Notably, all five isolates were sensitive to amoxicillin, chloramphenicol, penicillin, and tetracycline and exhibited resistance only to gentamicin. Considering their favourable probiotic properties, the selected LAB isolates were deemed potential candidates for use as probiotics in animal feed.

Supplementary Materials: The following supporting information can be downloaded at: https://www.mdpi.com/article/10.3390/agriculture14060924/s1, Table S1: Cycler program used for colony PCR with 27F and 1492R primers to amplify the 16S rRNA gene of bacterial isolates; Table S2: Representative sequences of the five isolates generated by Sanger sequencing using 27F and 1492R primers and their corresponding nucleotide BLAST results for taxonomic identification.

Author Contributions: Conceptualization, S.S.; methodology, S.S.; validation, S.S.; formal analysis, S.S., S.C., S.H. and T.K.; investigation, S.S., S.C., S.H. and T.K.; resources, S.S. and T.K.; writing—original draft preparation, S.S.; writing—review and editing, S.H. and T.K.; visualization, S.S. and T.K.; supervision, S.S.; project administration, S.S.; funding acquisition, S.S. and T.K. All authors have read and agreed to the published version of the manuscript.

Funding: This research project was financially supported by 'Thailand Science Research and Innovation (TSRI)'. Moreover, the research reported in this publication was jointly supported by the ASEAN-European Academic University Network (ASEA-UNINET), the Austrian Federal Ministry of Education, Science and Research (BMBWF), and the OeAD—Austria's Agency for Education and Internationalisation.

Institutional Review Board Statement: The research conducted in this study involved the use of *B. mori*, a widely cultivated model insect globally. As such, the study did not raise concerns related to experimental animal welfare, and no additional ethical approval was required for the procedures followed in this research.

Data Availability Statement: The 16S rRNA gene nucleotide sequences of the identified isolates have been publicly deposited in GenBank (https://www.ncbi.nlm.nih.gov/genbank/ (accessed on 20 May 2024)) under submission SUB14404462, covering accession numbers PP727389-PP727398.

Acknowledgments: The authors express their gratitude to the two anonymous reviewers for their valuable feedback, which significantly enhanced the manuscript.

Conflicts of Interest: The authors declare no conflicts of interest. The funders had no role in the design of the study; in the collection, analyses, or interpretation of data; in the writing of the manuscript; or in the decision to publish the results.

References

1. Alexandratos, N.; Bruinsma, J.; Alexandratos, N.; Bruinsma, J. *World Agriculture towards 2030/2050: The 2012 Revision*; Agricultural Development Economics Division, Food and Agriculture Organization of the United Nations: Rome, Italy, 2012. [CrossRef]
2. Zhou, Y.; Xiao, R.; Klammsteiner, T.; Kong, X.; Yan, B.; Mihai, F.-C.; Liu, T.; Zhang, Z.; Kumar Awasthi, M. Recent Trends and Advances in Composting and Vermicomposting Technologies: A Review. *Bioresour. Technol.* **2022**, *360*, 127591. [CrossRef] [PubMed]
3. Hanboonsong, Y.; Jamjanya, T.; Durst, P.B. *Six-Legged Livestock: Edible Insect Farming, Collection and Marketing in Thailand*; RAP Publication; Food and Agriculture Organization of the United Nations, Regional Office for Asia and the Pacific Bangkok, Thailand, 2013; ISBN 978-92-5-107578-4.
4. Orihara, Y.; Hamamoto, H.; Kasuga, H.; Shimada, T.; Kawaguchi, Y.; Sekimizu, K. A Silkworm–Baculovirus Model for Assessing the Therapeutic Effects of Antiviral Compounds: Characterization and Application to the Isolation of Antivirals from Traditional Medicines. *J. Gen. Virol.* **2008**, *89*, 188–194. [CrossRef] [PubMed]
5. Hamamoto, H.; Kurokawa, K.; Kaito, C.; Kamura, K.; Manitra Razanajatovo, I.; Kusuhara, H.; Santa, T.; Sekimizu, K. Quantitative Evaluation of the Therapeutic Effects of Antibiotics Using Silkworms Infected with Human Pathogenic Microorganisms. *Antimicrob. Agents Chemother.* **2004**, *48*, 774–779. [CrossRef] [PubMed]

6. Zhou, Y.; Zhou, S.; Duan, H.; Wang, J.; Yan, W. Silkworm Pupae: A Functional Food with Health Benefits for Humans. *Foods* **2022**, *11*, 1594. [CrossRef] [PubMed]
7. Banday, M.T.; Adil, S.; Sheikh, I.U.; Hamadani, H.; Qadri, F.I.; Sahfi, M.E.; Sait, H.S.A.W.; Abd El-Mageed, T.A.; Salem, H.M.; Taha, A.E.; et al. The Use of Silkworm Pupae (*Bombyx mori*) Meal as an Alternative Protein Source for Poultry. *Worlds Poult. Sci. J.* **2023**, *79*, 119–134. [CrossRef]
8. Park, J.-H.; Lee, D.-G.; Yeon, S.-W.; Kwon, H.-S.; Ko, J.-H.; Shin, D.-J.; Park, H.-S.; Kim, Y.-S.; Bang, M.-H.; Baek, N.-I. Isolation of Megastigmane Sesquiterpenes from the Silkworm (*Bombyx mori* L.) Droppings and Their Promotion Activity on HO-1 and SIRT1. *Arch. Pharm. Res.* **2011**, *34*, 533–542. [CrossRef] [PubMed]
9. Łochyńska, M.; Frankowski, J. The Biogas Production Potential from Silkworm Waste. *Waste Manag.* **2018**, *79*, 564–570. [CrossRef]
10. Łochyńska, M.; Frankowski, J. Impact of Silkworm Excrement Organic Fertilizer on Hemp Biomass Yield and Composition. *J. Ecol. Eng.* **2019**, *20*, 63–71. [CrossRef]
11. Van De Zande, E.M.; Wantulla, M.; Van Loon, J.J.A.; Dicke, M. Soil Amendment with Insect Frass and Exuviae Affects Rhizosphere Bacterial Community, Shoot Growth and Carbon/Nitrogen Ratio of a Brassicaceous Plant. *Plant Soil* **2023**, *495*, 631–648. [CrossRef]
12. Vimolmangkang, S.; Somkhanngoen, C.; Sukrong, S. Potential Pharmaceutical Uses of the Isolated Compounds from Silkworm Excreta. *Chiang Mai J. Sci.* **2013**, *41*, 97.
13. Tulp, M.; Bohlin, L. Unconventional Natural Sources for Future Drug Discovery. *Drug Discov. Today* **2004**, *9*, 450–458. [CrossRef] [PubMed]
14. Huang, X.; Wu, J.; Li, Q.; Gu, S.; Chen, Z.; Wang, K. Silkworm Feces Extract Improves Iron Deficiency Anemia via Suppressing Hepcidin Expression and Promoting Iron-Regulatory Proteins Expression. *RSC Adv.* **2017**, *7*, 50378–50388. [CrossRef]
15. Neelagund, S.; Ingalhalli, S.; Savanurmath, C.; Hinchigeri, S.; Hiremath, M. Purification and Characterization of Antiviral Protein from Silkworm Fecal Matter. *Casp. J. Environ. Sci.* **2007**, *5*, 77–85.
16. Qin, D.; Wang, G.; Dong, Z.; Xia, Q.; Zhao, P. Comparative Fecal Metabolomes of Silkworms Being Fed Mulberry Leaf and Artificial Diet. *Insects* **2020**, *11*, 851. [CrossRef] [PubMed]
17. Huerta-García, A.; Álvarez-Cervantes, J. The Gut Microbiota of Insects: A Potential Source of Bacteria and Metabolites. *Int. J. Trop. Insect Sci.* **2024**, *44*, 13–30. [CrossRef]
18. Liang, X.; Fu, Y.; Liu, H. Isolation and Characterization of Enzyme-Producing Bacteria of the Silkworm Larval Gut in Bioregenerative Life Support System. *Acta Astronaut.* **2015**, *116*, 247–253. [CrossRef]
19. Unban, K.; Klongklaew, A.; Kodchasee, P.; Pamueangmun, P.; Shetty, K.; Khanongnuch, C. Enterococci as Dominant Xylose Utilizing Lactic Acid Bacteria in Eri Silkworm Midgut and the Potential Use of *Enterococcus hirae* as Probiotic for Eri Culture. *Insects* **2022**, *13*, 136. [CrossRef]
20. Yeruva, T.; Vankadara, S.; Ramasamy, S.; Lingaiah, K. Identification of Potential Probiotics in the Midgut of Mulberry Silkworm, *Bombyx mori* through Metagenomic Approach. *Probiotics Antimicrob. Proteins* **2020**, *12*, 635–640. [CrossRef] [PubMed]
21. Dong, H.-L.; Zhang, S.-X.; Chen, Z.-H.; Tao, H.; Li, X.; Qiu, J.-F.; Cui, W.-Z.; Sima, Y.-H.; Cui, W.-Z.; Xu, S.-Q. Differences in Gut Microbiota between Silkworms (*Bombyx mori*) Reared on Fresh Mulberry (*Morus alba* Var. Multicaulis) Leaves or an Artificial Diet. *RSC Adv.* **2018**, *8*, 26188–26200. [CrossRef]
22. FAO/WHO. *Guidelines for the Evaluation of Probiotics in Food;* Safety Evaluation of Certain Contaminants in Food; Food and Agriculture Organization: Rome, Italy, 2006.
23. Bermudez-Brito, M.; Plaza-Díaz, J.; Muñoz-Quezada, S.; Gómez-Llorente, C.; Gil, A. Probiotic Mechanisms of Action. *Ann. Nutr. Metab.* **2012**, *61*, 160–174. [CrossRef]
24. Guarner, F. Probiotics. *Int. J. Food Microbiol.* **1998**, *39*, 237–238. [CrossRef] [PubMed]
25. Lähteinen, T.; Malinen, E.; Koort, J.M.K.; Mertaniemi-Hannus, U.; Hankimo, T.; Karikoski, N.; Pakkanen, S.; Laine, H.; Sillanpää, H.; Söderholm, H.; et al. Probiotic Properties of *Lactobacillus* Isolates Originating from Porcine Intestine and Feces. *Anaerobe* **2010**, *16*, 293–300. [CrossRef] [PubMed]
26. Jiao, Y.; Yang, H.; Shigwedha, N.; Zhang, S.; Liu, F.; Zhang, L. Probiotic Effects and Metabolic Products of *Enterococcus faecalis* LD33 with Respiration Capacity. *Foods* **2022**, *11*, 606. [CrossRef] [PubMed]
27. Al Atya, A.K.; Drider-Hadiouche, K.; Ravallec, R.; Silvain, A.; Vachee, A.; Drider, D. Probiotic Potential of *Enterococcus faecalis* Strains Isolated from Meconium. *Front. Microbiol.* **2015**, *6*, 227. [CrossRef] [PubMed]
28. Liu, Z.; Chen, Y.; Meng, Q.; Zhang, X.; Wang, X. Progress in the Application of *Enterococcus faecium* in Animal Husbandry. *Front. Cell. Infect. Microbiol.* **2023**, *13*, 1168189. [CrossRef] [PubMed]
29. Jena, P.K.; Trivedi, D.; Thakore, K.; Chaudhary, H.; Giri, S.S.; Seshadri, S. Isolation and Characterization of Probiotic Properties of Lactobacilli Isolated from Rat Fecal Microbiota. *Microbiol. Immunol.* **2013**, *57*, 407–416. [CrossRef] [PubMed]
30. Dobrovolskaia, M.A.; Clogston, J.D.; Neun, B.W.; Hall, J.B.; Patri, A.K.; McNeil, S.E. Method for Analysis of Nanoparticle Hemolytic Properties in Vitro. *Nano Lett.* **2008**, *8*, 2180–2187. [CrossRef]
31. Mangia, N.P.; Saliba, L.; Deiana, P. Functional and Safety Characterization of Autochthonous *Lactobacillus paracasei* FS103 Isolated from Sheep Cheese and Its Survival in Sheep and Cow Fermented Milks during Cold Storage. *Ann. Microbiol.* **2019**, *69*, 161–170. [CrossRef]
32. Zhao, L.; Zhang, Y.; Liu, Y.; Zhong, J.; Zhang, D. Assessing the Safety and Probiotic Characteristics of *Lacticaseibacillus rhamnosus* X253 via Complete Genome and Phenotype Analysis. *Microorganisms* **2023**, *11*, 140. [CrossRef]

33. Boiocchi, F.; Porcellato, D.; Limonta, L.; Picozzi, C.; Vigentini, I.; Locatelli, D.P.; Foschino, R. Insect Frass in Stored Cereal Products as a Potential Source of *Lactobacillus sanfranciscensis* for Sourdough Ecosystem. *J. Appl. Microbiol.* **2017**, *123*, 944–955. [CrossRef]
34. Hamidreza, T.; Fatemeh, T.; Hossein, M.; Mojtaba, Z.; Mahmood, S.; Parvin, S. Potential Probiotic of *Lactobacillus johnsonii* LT171 for Chicken Nutrition. *Afr. J. Biotechnol.* **2009**, *8*, 5833–5837. [CrossRef]
35. Krausova, G.; Hyrslova, I.; Hynstova, I. In Vitro Evaluation of Adhesion Capacity, Hydrophobicity, and Auto-Aggregation of Newly Isolated Potential Probiotic Strains. *Fermentation* **2019**, *5*, 100. [CrossRef]
36. Schillinger, U.; Lücke, F.K. Antibacterial Activity of *Lactobacillus sake* Isolated from Meat. *Appl. Environ. Microbiol.* **1989**, *55*, 1901–1906. [CrossRef]
37. Vélez, M.P.; Hermans, K.; Verhoeven, T.L.A.; Lebeer, S.E.; Vanderleyden, J.; De Keersmaecker, S.C.J. Identification and Characterization of Starter Lactic Acid Bacteria and Probiotics from Columbian Dairy Products. *J. Appl. Microbiol.* **2007**, *103*, 666–674. [CrossRef]
38. Bauer, A.W.; Kirby, W.M.M.; Sherris, J.C.; Turck, M. Antibiotic Susceptibility Testing by a Standardized Single Disk Method. *Am. J. Clin. Pathol.* **1966**, *45*, 493–496. [CrossRef]
39. Marchwińska, K.; Gwiazdowska, D. Isolation and Probiotic Potential of Lactic Acid Bacteria from Swine Feces for Feed Additive Composition. *Arch. Microbiol.* **2022**, *204*, 61. [CrossRef]
40. Pérez-Sánchez, T.; Balcázar, J.L.; García, Y.; Halaihel, N.; Vendrell, D.; De Blas, I.; Merrifield, D.L.; Ruiz-Zarzuela, I. Identification and Characterization of Lactic Acid Bacteria Isolated from Rainbow Trout, *Oncorhynchus mykiss* (Walbaum), with Inhibitory Activity against *Lactococcus garvieae*: Trout Endogenous LAB Antagonise *L. garvieae*. *J. Fish Dis.* **2011**, *34*, 499–507. [CrossRef]
41. Dhakal, R.; Sahu, R.K.; Das Gupta, D.; Saikia, M.; Borthakur, S.; Majumder, M.; Mech, S.; Thapa, M.; Dutta, P.; Kalita, J. Recycling of Protein Rich Silk Industry Waste for Potential Food and Therapeutic Application. *Food Biosci.* **2024**, *60*, 104461. [CrossRef]
42. Praeg, N.; Klammsteiner, T. Primary study on frass fertilizers from mass-reared insects: Species variation, heat treatment effects, and implications for soil application at laboratory scale. *J. Environ. Manag.* **2024**, *356*, 120622. [CrossRef]
43. Insam, H.; Klammsteiner, T.; Gómez-Brandòn, M. Biology of Compost. In *Encyclopedia of Soils in the Environment*; Elsevier: Amsterdam, The Netherlands, 2023; pp. 522–532. ISBN 978-0-12-409548-9.
44. Menconi, A.; Kallapura, G.; Latorre, J.D.; Morgan, M.J.; Pumford, N.R.; Hargis, B.M.; Tellez, G. Identification and Characterization of Lactic Acid Bacteria in a Commercial Probiotic Culture. *Biosci. Microbiota Food Health* **2014**, *33*, 25–30. [CrossRef]
45. Jaimee, G.; Halami, P.M. Emerging Resistance to Aminoglycosides in Lactic Acid Bacteria of Food Origin—An Impending Menace. *Appl. Microbiol. Biotechnol.* **2016**, *100*, 1137–1151. [CrossRef] [PubMed]
46. Klibi, N.; Aouini, R.; Borgo, F.; Ben Said, L.; Ferrario, C.; Dziri, R.; Boudabous, A.; Torres, C.; Ben Slama, K. Antibiotic Resistance and Virulence of Faecal Enterococci Isolated from Food-Producing Animals in Tunisia. *Ann. Microbiol.* **2015**, *65*, 695–702. [CrossRef]
47. Zhang, X.; Feng, H.; He, J.; Muhammad, A.; Zhang, F.; Lu, X. Features and Colonization Strategies of *Enterococcus faecalis* in the Gut of *Bombyx mori*. *Front. Microbiol.* **2022**, *13*, 921330. [CrossRef] [PubMed]
48. Zhang, X.; Feng, H.; He, J.; Liang, X.; Zhang, N.; Shao, Y.; Zhang, F.; Lu, X. The Gut Commensal Bacterium *Enterococcus faecalis* LX10 Contributes to Defending against *Nosema bombycis* Infection in *Bombyx mori*. *Pest Manag. Sci.* **2022**, *78*, 2215–2227. [CrossRef]
49. Torres, C.; Alonso, C.A.; Ruiz-Ripa, L.; León-Sampedro, R.; Del Campo, R.; Coque, T.M. Antimicrobial Resistance in *Enterococcus* spp. of Animal Origin. *Microbiol. Spectr.* **2018**, *6*, 4. [CrossRef] [PubMed]
50. Sun, Z.; Lu, Y.; Zhang, H.; Kumar, D.; Liu, B.; Gong, Y.; Zhu, M.; Zhu, L.; Liang, Z.; Kuang, S.; et al. Effects of BmCPV Infection on Silkworm *Bombyx mori* Intestinal Bacteria. *PLoS ONE* **2016**, *11*, e0146313. [CrossRef]
51. Cappellozza, S.; Saviane, A.; Tettamanti, G.; Squadrin, M.; Vendramin, E.; Paolucci, P.; Franzetti, E.; Squartini, A. Identification of *Enterococcus mundtii* as a Pathogenic Agent Involved in the "Flacherie" Disease in *Bombyx mori* L. Larvae Reared on Artificial Diet. *J. Invertebr. Pathol.* **2011**, *106*, 386–393. [CrossRef]
52. Grau, T.; Vilcinskas, A.; Joop, G. Probiotic *Enterococcus mundtii* Isolate Protects the Model Insect *Tribolium castaneum* against *Bacillus thuringiensis*. *Front. Microbiol.* **2017**, *8*, 1261. [CrossRef]
53. Cebeci, A.; Gürakan, C. Properties of Potential Probiotic *Lactobacillus plantarum* Strains. *Food Microbiol.* **2003**, *20*, 511–518. [CrossRef]
54. Ripamonti, B.; Agazzi, A.; Bersani, C.; De Dea, P.; Pecorini, C.; Pirani, S.; Rebucci, R.; Savoini, G.; Stella, S.; Stenico, A.; et al. Screening of Species-Specific Lactic Acid Bacteria for Veal Calves Multi-Strain Probiotic Adjuncts. *Anaerobe* **2011**, *17*, 97–105. [CrossRef]
55. Succi, M.; Tremonte, P.; Reale, A.; Sorrentino, E.; Grazia, L.; Pacifico, S.; Coppola, R. Bile Salt and Acid Tolerance of *Lactobacillus rhamnosus* Strains Isolated from Parmigiano Reggiano Cheese. *FEMS Microbiol. Lett.* **2005**, *244*, 129–137. [CrossRef] [PubMed]
56. Bazireh, H.; Shariati, P.; Azimzadeh Jamalkandi, S.; Ahmadi, A.; Boroumand, M.A. Isolation of Novel Probiotic *Lactobacillus* and *Enterococcus* Strains from Human Salivary and Fecal Sources. *Front. Microbiol.* **2020**, *11*, 597946. [CrossRef] [PubMed]
57. Del Re, B.; Sgorbati, B.; Miglioli, M.; Palenzona, D. Adhesion, Autoaggregation and Hydrophobicity of 13 Strains of *Bifidobacterium longum*. *Lett. Appl. Microbiol.* **2000**, *31*, 438–442. [CrossRef]
58. Patel, A.K.; Ahire, J.J.; Pawar, S.P.; Chaudhari, B.L.; Chincholkar, S.B. Comparative Accounts of Probiotic Characteristics of *Bacillus* spp. Isolated from Food Wastes. *Food Res. Int.* **2009**, *42*, 505–510. [CrossRef]

59. Asadi, A.; Lohrasbi, V.; Abdi, M.; Mirkalantari, S.; Esghaei, M.; Kashanian, M.; Oshaghi, M.; Talebi, M. The Probiotic Properties and Potential of Vaginal *Lactobacillus* spp. Isolated from Healthy Women against Some Vaginal Pathogens. *Lett. Appl. Microbiol.* **2022**, *74*, 752–764. [CrossRef]
60. Krawczyk, B.; Wityk, P.; Gałęcka, M.; Michalik, M. The Many Faces of *Enterococcus* spp.—Commensal, Probiotic and Opportunistic Pathogen. *Microorganisms* **2021**, *9*, 1900. [CrossRef] [PubMed]
61. Servin, A.L. Antagonistic Activities of Lactobacilli and Bifidobacteria against Microbial Pathogens. *FEMS Microbiol. Rev.* **2004**, *28*, 405–440. [CrossRef]
62. Savio, C.; Mugo-Kamiri, L.; Upfold, J.K. Bugs in Bugs: The Role of Probiotics and Prebiotics in Maintenance of Health in Mass-Reared Insects. *Insects* **2022**, *13*, 376. [CrossRef]

agriculture

MDPI

Article

A Multidisciplinary Approach for the Development of a Supply Chain in Biomass Conversion of Agrifood Waste Mediated by Larvae of *Hermetia illucens* L.: From Rearing to By-Product Exploitation

Eleonora De Santis [1,2], Alberto de Iudicibus [1,3], Francesca Lecce [4], Massimiliano De Mei [2], Francesco Petrazzuolo [2], Angelo Del Giudice [1], Monica Carnevale [1], Francesco Gallucci [1], Claudio Beni [1], Alberto Assirelli [1], Enrico Santangelo [1,*] and Silvia Arnone [2,*]

1 Council for Agricultural Research and Economics (CREA), Research Centre for Engineering and Agro-Food Processing, Via della Pascolare 16, 00015 Monterotondo, Italy; eledes2106@gmail.com (E.D S.); alberto.deiudicibus@crea.gov.it (A.d.I.); francesco.gallucci@crea.gov.it (F.G.)

2 Energy Technologies and Renewable Sources Department, Italian National Agency for New Technologies, Energy and Sustainable Economic Development (ENEA), Via Anguillarese 301, 00123 Santa Maria di Galeria, Italy; massimiliano.demei@enea.it (M.D.M.)

3 Council for Agricultural Research and Economics (CREA), Research Centre for Plant Protection and Certification, Via C. G. Bertero 22, 00156 Rome, Italy

4 Department of Sustainability, Italian National Agency for New Technologies, Energy and Sustainable Economic Development (ENEA), Via Anguillarese 301, 00123 Santa Maria di Galeria, Italy

* Correspondence: enrico.santangelo@crea.gov.it (E.S.); silvia.arnone@enea.it (S.A.); Tel.: +39-0630484656 (S.A.)

Abstract: Black soldier fly larvae (BSFL) can convert various organic substrates into high added-value biomass. In addition, the residue can be used as a soil conditioner. Several studies have been conducted on a laboratory scale that may not represent what happens on a prototype scale. Using fruit and vegetable waste as a basic substrate, mixing them with agro-industry by-products (called co-substrates), the Hermes project set up a process on medium (2 kg) and large (10 kg) scales with two different feeding regimes (1.25 g/BSFL and 2 g/BSFL). At the mature stage, larval biomass was separated from frass (the by-product of the larval rearing). The production of larval proteins and fats and the use of frass as soil conditioning were evaluated. The lowest feeding regime (1.25 g/BSFL) provided the best waste valorization. The shift towards higher production scales is not completely linear. The addition of co-substrates to fruit and vegetable waste, as they are provided by the large-scale retail trade, can help to standardize a process as part of an insect farm. The frass recovered from the residue of rearing (on the diet or on the agrifood leftovers) was composted and used in field to grow a processing tomato variety. The addition of composted frass assured a slightly lower yield than synthetic fertilizer but there was no statistically significant difference ($p > 0.10$). This suggests that partial replacement of synthetic fertilizer with composted frass has potential. Overall, the work demonstrated that, using a multidisciplinary approach, the interest and the value in building a supply chain based on bioconversion mediated by *Hermetia illucens* can be emphasized.

Keywords: circular economy; waste reduction; bioconversion; CORS; *Hermetia illucens*; frass; organic fertilizers

1. Introduction

By 2050, the world population is estimated to grow to about 9.6 billion and the demand for animal protein to increase by 70% [1]. With this rapid population growth, economic development, and urbanization, meeting the demand for feed and quality food and organic waste management require substantial resources, often with great environmental impact [2].

The European Commission highlights organic waste as a key item of the circular and green economy, emphasizing its valorization into useful products. Several studies investigated their potential added value, mainly using anaerobic digestion and composting [3]. Among the organic biomass, fruit and vegetable waste (FVW), a mixture of edible and inedible leftovers, plays an important role. The Food and Agriculture Organization (FAO) reports about 45% of fruits and vegetables (FV) are discarded (as waste) across the food supply chain [2]. Furthermore, the United Nations predicts that 13% of all agricultural product is lost during harvest as well as retail, while 17% is wasted in houses, retail, and eating places [4]. In the context of FVW valorization, one promising strategy is its exploitation as a substrate for mass rearing of edible insects to be used as a protein source for aquaculture, chicken farming, or as a lipid source for biodiesel production [5,6].

Saprophagous insects represent a potentially valid solution to two problems: the recovery of the growing amount of organic waste and the increasing global demand for feed and food. In a circular economy perspective, the application of Conversion of Organic Refuse by Saprophages (CORS) technology mediated by the black soldier fly (BSF), *Hermetia illucens* L. (Diptera, Stratiomyidae, Hermetiinae), represents an important innovation that can significantly contribute to the development of supply chains that meet the principles of green economy and green chemistry. The species is distributed globally throughout temperate and tropical regions [7]. BSF larvae (BSFL) can grow on various organic substrates, i.e., vegetable matter, manure, and catering waste, transforming them into high added-value biomass. It was observed that mature larvae, fed on different organic waste, contain up to 44% of dry matter and reach 42% of protein and 35% of fat [8–10].

Larval performances, in terms of growth, size, biochemical composition, and life span, as well as the efficiency by which substrates are converted into larval biomass, vary considerably and are affected by the quality of the substrate [11–13], substrate supply [14], as well as temperature, oxygen availability, and moisture content [15,16].

Larvae can grow on substrate with a moisture content within the range of 60–90% [17], with the optimal at 70–75% [15]. For these reasons, the use of FVW as BSFL growth substrate may not be sufficient to create suitable conditions for the optimal larval development and the bioconversion of the waste into high added-value molecules, due to high moisture content of the substrate. Furthermore, in case of high moisture content of such a waste stream, the separation of BSFL from the frass might be challenging and time consuming [18].

Several works have investigated the feasibility of mixing FVW with other types of waste that would optimize the bioconversion process, like fermented "spent coffee grounds" or bread waste from catering [19,20]. Besides abiotic factors, such as substrate quality, temperature, and moisture content, the feeding regime intended as substrate amount per larva (sometimes stated as larval density, i.e., number of larvae per area unit) is an important factor that needs to be optimized for its ability to improve or hinder the BSFL growth [14,21]. At low larval densities, which means a great amount of substrate per larva, BSFL may not achieve sufficient substrate conditioning and co-operative digestion, while too high densities lead to providing a low quantity of feed per larva, causing competition over the nutrients, and may negatively influence the growth [22]. The density also influences other physical phenomena such as heat storage and evaporation, which change the feed properties and consequently affect the growth of the larvae [23].

The bioconversion process reduces the volume by 50–70% of waste, a result that could alleviate the environmental and economic cost of their disposal [24]. At the end of the bioconversion cycle, the larval biomass can be separated and used directly as feed or as raw material in a biorefinery conversion to produce protein meal, oil, biodiesel, and chitin [25]. The rearing residue (frass) is a mixture of uneaten substrate, feces, and exuviae, which has interesting characteristics for improving soil and crop productivity [26]. Several works have tested the frass as soil conditioner as such or after composting, demonstrating its potential as alternative organic fertilizer on different crops [26–29].

Historically, most of the studies concerning the knowledge on BSF feeding performance on different waste streams have been conducted at a laboratory scale (i.e., several hundred larvae per replicate on hundred grams of substrate) showing how BSFL growth and performance can vary on different waste or with different bioconversion conditions (feeding rate, moisture of the substrate, larval density, and feeding regime). The shifting from laboratory to industrial scale is not necessarily linear; many authors indicated that the findings of small-scale trials could not be necessarily moved to large production systems. For example, small-scale compared to large-scale studies, both carried out in the same conditions on swine manure, gave different results for larval development time until the mature stage (less in the small scale) or survivorship (higher in the small scale) or for the prepupal weight (less in the small scale) [30]. The size of the pilot-level studies (also named large-scale or mass production or industrial-level studies) ranges from 10,000 to 20,000 larvae on 6–10 kg of substrate in a crate of about 30–40 L of volume, considering each crate as a bioconversion unit [31–34]. Gligorescu et al. [32] found the best feeding treatment testing the bioconversion efficiency in crates with 20,000 BSFL on 8–10 kg of food waste. Yang-Jie et al. [35] planned a 4-year full-scale study with containers filled with 6 kg of urban waste inoculated with 8000 BSFL. The containers were racked in rows and layers to treat 15 ton/day of waste. Over the 4 years, they could improve the production of larvae and the protein content modifying the frequency of pile turning or changing the fermentation agent of the initial substrate. Hence, the industrial bioconversion process must be developed by continuous technical optimization, according to the waste resource (type and quantity daily provided).

The HERMES project, funded within POR FESR Lazio 2014–2020 (Det. Reg. n. G09493 140721, 22/07/2021), aimed at making available a sustainable and environmentally friendly technology for the recovery and valorization of FVW from the large-scale retail trade by means of bioconversion mediated by BSFL. With a view of the likely seasonal variability of FVW, it was planned to carry out the bioconversion trials using waste provided by one supermarket and, since the large-scale tests required a longer time for the preparation and more space than the medium-scale laboratory tests, trials were diluted over time, within the time allowed for the completion of the Hermes project. Hence, a supermarket provided the mixture of FVW all over the year. To lower the moisture content of the FVW substrate, bakery waste bran, brewer's spent grain, and dry stuff (such as cellulose ramekins and straw) were added one at a time in different tests. All the by-products were chosen since they were locally available.

The experiments were planned in order to (i) compare two feeding regimes and two rearing scales on bioconversion efficiency parameters (larval biomass, added-value products, and BSFL waste reduction); (ii) verify the scale-up of the system from experimental proof-of-concept on a small-scale lab environment to the large scale; and (iii) evaluate the field behavior of the composted residue.

2. Materials and Methods

2.1. Insect Rearing

Specimens for the experiments were obtained from the *H. illucens* colony, established at the ENEA Casaccia Research Center of Rome, based on the rearing process described by Harnden & Tomberlin [36], which is a modified method of Sheppard et al. [37].

Adults were confined in a BugDorm cage (type 6M630) (60 cm × 60 cm × 180 cm), placed inside a climatic chamber maintained at 27 ± 1 °C, 70 ± 10% relative humidity (RH), and 16:8 (light/dark) photoperiod controlled by BEF Biosystems LED lights. The following were placed inside the cage: a Petri dish containing absorbent cotton soaked in a sugar-water solution, a beaker containing only water, and the oviposition site consisting of a beaker on the bottom of which "spent" diet, separated by a mesh, was placed to attract females. Strips of corrugated cardboard placed on the beaker's top edge were used by the flies for oviposition. Eggs were collected after 24 h exposure to oviposing females.

BSFL were fed on a standard Gainesville diet, consisting of 20% corn, 30% alfalfa meal, and 50% wheat bran, mixed with water in a 1:1.7 ratio (100 g diet:170 mL water) [38]. The larvae were reared in a climatic cabinet maintained at a temperature of 27 ± 1 °C,

70 ± 10% RH, and darkness. The rearing was carried out adopting a "semi-batch system", i.e., adding the substrate (~0.1 g/BSFL/day) necessary to complete the BSFL growing (about 1300 g/1000 BSFL) three times (instead of once at the beginning of the batch system). Once larvae reached the mature stage, they were separated from the residue and transferred into a new climatic chamber at 27 ± 1 °C, 50 ± 10% RH, and darkness for pupation. The frass of the rearing on the Gainesville diet was stored in the freezer until being used for composting tests and was named GD_{comp}.

2.2. The Substrates and the Experimental Trials

The basic substrate was a mixture of fruit and vegetable waste (FVW) provided by the Unicoop Tirreno supermarket located in Rome. With a view of the likely seasonal variability of FVW, it was planned to carry out the bioconversion trials by covering the entire year with waste collected in the spring–summer and fall–winter periods. For this purpose, an agreement was established with Unicoop for the continuing supply. As was expected, the composition of the basic substrate FVW varied with every supply. This variation has not been considered as a variable to follow a real-life scenario of a daily supply to a bioconversion plant. The parameter that addressed the choice about the addition of other residual biomass was the percentage of humidity of the total FVW. Since optimal larval growth requires 70–75% moisture content of the feeding substrate [15], different co-substrates have been added to lower the overabundant moisture content of FVW (95–98%). The bakery waste (BW) of the same supermarket has been firstly considered and was used for the first trial of this work (trial A). As a second step, bran (BR) (from a local seller), BSG (from two artisanal breweries in the province of Rome), and dry stuff (DS), such as straw or cellulose ramekins, were considered for a second experiment (trial B). Based on their moisture content, the residual biomasses were added to FVW at the following ratios: 80:20 (FVW:BW and BR), 60:40 (FVW:BSG), and 98:2 (FVW:DS). FVW and BD were chopped by hand before feeding the larvae; FVW, BW, BR, and BSG were carefully mixed; DS was placed in a layer of about 1–2 cm at the bottom of the box.

Table 1 shows the configuration of experiments A and B, in terms of scale (medium scale, MS, and large scale, LS), feeding regime (low feeding regime, LFR, and high feeding regime, HFR), and type of the substrate (1, 2, 3, and 4).

Table 1. The configuration of Trial A and Trial B.

Trial	Scale	Larval Feeding Regime	Subst.	Waste (%)				
					FW [1]		VW [2]	BW [3]
A	MS [4]	HFR [5]	1	.5	30		50	20
		LFR [6]	1		30		50	20
	LS [7]	HFR	1		20		60	20
		LFR	1		20		60	20
				FW	VW	BR [8]	BSG [9]	DS [10]
B	MS	LFR	2		80	20		
			3	10	50		40	
			4		98			2
	LS	LFR	2	30	50	20		
			3	10	50		40	
			4	9	89			2

[1] FW, fruit waste; [2] VW, vegetable waste; [3] BW, bakery waste; [4] MS, medium scale (2 kg of substrate); [5] HFR, high feeding regime (2 g of substrate/BSFL); [6] LFR, low feeding regime (1.25 g of substrate/BSFL); [7] LS, large scale (10 kg of substrate); [8] BR, bran; [9] BSG, brewer's spent grains; [10] DS, dry stuff.

- Trial A was aimed to evaluate substrate 1 (FVW:BW), comparing two feeding regimes intended as the total amount of substrate divided per total number of larvae; the HFR consisted of 2 g/BSFL and the LFR of 1.25 g/BSFL, both applied at MS (2 kg of substrate) and LS (10 kg of substrate). At the end of the experiment, the effective feeding rates (g/BSFL/day) were calculated for each experiment, dividing the amount of provided substrate per number of days until larvae reached a mature stage.
- Trial B was aimed to evaluate the linearity of bioconversion efficiency in the shift towards higher production scales, comparing MS with LS trials. On the basis of the results of trial A, tests were carried out only adopting the LFR and were conducted using the other three substrate mixtures: FVW:BR (2), FVW:BSG (3), and FVW:DS (4). The data obtained in trial A with substrate 1 (MS- and LS-LFR1) were compared with data obtained with substrates 2, 3, and 4.

Plastic boxes of 36.5 cm × 25.5 cm × 14 H were used for MS trials; plastic boxes of 59 cm × 39 cm × 17 H were used for LS trials. All the boxes were covered with perforated lids lined with a mesh of 1 mm^2. For all the trials, 6-day-old larvae reared until then on a Gainesville diet were used and exposed to a batch feeding strategy. The experimental trials, carried out in triplicate, were kept in a dark climate chamber maintained at 27 ± 1 °C and 70 ± 10% RH.

2.3. *Bioconversion Efficiency Parameters*

Larvae were weighed at the beginning of the experiment and every 2–3 days during the trial. Three randomized subsamples of 30 larvae each were washed, dried on filter paper, weighed, and then returned to their respective container. The trials ended when the larval weight began to decrease. At this phase, mature larvae undertake the prepupal period with a nonfeeding phase, accompanied by weight loss and by darkening of the body color [34,39]. Once the larvae reached the prepupal stage, the larval biomass was separated from the residue (frass) that was stored in the freezer until being used for composting tests and was named FVW_{comp}.

Larvae were washed, dried on filter paper, and then weighed. Larvae and substrates were characterized as follows: MLW was based on fresh weight; dry matter, ash, protein, and lipid content of both initial substrate and mature larvae were determined using standard procedures [40]. Before starting each experiment, 100 g of fresh substrate was oven-dried at 105 °C for 24 h to determine dry matter. At the end of the bioconversion, 40 g of fresh larval biomass was boiled for 3 min [41] and then oven-dried at 60 °C for 48 h to determine dry matter. Ash content was determined in a muffle furnace at 550 °C for 4 h for all the samples. Crude fat content was determined on dry sample by means of an automatized Soxhlet extractor (Soxhtraction PBI International apparatus, Milano, Italy), using *n*-hexane as solvent [42]. Total nitrogen was determined on the defatted sample with the Kjeldahl system (Buchi 426 Digestion Unit and Buchi 323 Distillation Unit, Cornaredo, Italy) [43]. To determine protein content, total N was multiplied with the factor Kp 6.25 for the substrate and Kp 4.75 for the larval biomass to avoid an overestimation of proteins due to the presence of chitin in the exoskeleton [44]. All analyses were performed in triplicate.

Carbohydrate content (%) was estimated by difference, subtracting the sum of lipids, protein, and ash (g/100 gDM) from the total dry weight of the sample (100 gDM):

$$\text{Carbohydrates} = 100 - [(\text{crude lipid} + \text{crude protein} + \text{ash})\ \text{content}\ (\text{g}/100\ \text{g})] \quad (1)$$

Protein conversion ratio (% PrCR) was calculated on a dry matter basis following Lalander et al. [34]:

$$\text{PrCR} = \frac{\text{Pr}_{\text{BSFL}}}{\text{DM}_{\text{BSFL}}} \times \frac{\text{Pr}_{\text{IS}}}{\text{DM}_{\text{IS}}} \times 100 \quad (2)$$

where DM_{BSFL} and DM_{IS} are the total dry matter in the mature larvae and initial substrate, respectively, while Pr_{BSFL} and Pr_{IS} are the percentage of crude protein (% DM) in the mature larvae and initial substrate, respectively.

Waste reduction (% WR) was calculated on a wet weight basis as follows:

$$WR = \frac{\text{Initial substrate (g)} - \text{Residue (g)}}{\text{Initial substrate (g)}} \times 100 \quad (3)$$

2.4. Field Trial

In the case of an industrial plant processing heterogeneous material, as is the case of the present study, it is necessary to standardize the reproductive cycle of the insect using for larval development an artificial diet. The frass produced from the insect rearing starting from the Gainsville diet (GD) or from FVW were composted in two cylindrical composters (H 100, Ø 48 cm approx.) made of wire mesh covered with green shade cloth. Each composter had a capacity of approximately 25 kg and was intended for the dietary residue or the fruit and vegetable residue after bioconversion. Wheat straw was added to the frass in a 1:1 (w/w) ratio. The cylinders were irrigated for one hour periodically with a rotary sprinkler. The composting phase was conducted between the beginning of February and mid-June 2023, lasting for 100 days.

The final compost was analyzed for the content of the main elements and heavy metals. Before the characterization, 50 g of each substrate was dried for 24 h at 105 ± 2 °C (Memmert UFP800 drying oven, Schwabach, Germany) to determine the moisture content [45]. The dried biomass was treated with a Retsch SM 100 kneading mill and then with a centrifugal mill (Retsch ZM 200, Retsch, Haan, Germany) to shred and homogenize the matrix. To determine the content of carbon (C) and nitrogen (N), about 5 mg of each sample was analyzed by Costech ECS 4010 CHNS-O elemental analyzer (Costech International, Pioltello, Italy) [46]. The quantification limit (LOQ) for each sample was 0.05% w/w. For the macro- and trace element content, 0.5 g of the dehydrated sample was homogenized and solubilized with 6 ± 0.1 mL of HNO_3 65% and 3 ± 0.1 mL of H_2O_2 30%. The solution was digested in a microwave oven at 180 °C, 650 W, for 8 min and then for a further 15 min. At the end of digestion, the samples were filtered and diluted with MilliQ water. Two replicates and a blank were made for each sample. The calibration line was made in nitric acid on five points at increasing concentrations of internal standard. Element analysis was performed using ICP-MS (Agilent 7700, Agilent Technologies, Tokyo, Japan).

Tomato plantlets (*Solanum lycopersicum*, L.) of the Roma Nano F1 variety, at the 4–5th true leaf stage, were transplanted on May 2023 at the experimental field of the CREA Research Centre for Engineering and Agro-Food Processing, Monterotondo, Italy (42° N 05′056.86″, 12° E 37′026.23″). The soil, belonging to silty clayey loam USDA classification (Table 2), was left as fallow in the preceding year. In the recent past, the soil was fertilized with cow manure for at least three years.

Table 2. Soil physical and chemical properties.

Soil Properties	U.M. [1]	
Sand	%	25
Silt	%	38
Clay	%	37
pH		7.9
Organic matter	%	2.6
Total nitrogen (N)	%	0.16
Assimilable phosphorous (P)	mg/kg	14.1
Exchangeable potassium (K)	mg/kg	365.3
Cation Exchange Capacity	Meq/100 g	31.8

[1] Unit of measurement.

The soil composition was studied by Bascietto et al. [47], which outlined that the endowment in organic matter and total nitrogen (N) was moderate, while potassium (K) was well represented. Overall, N and K concentrations were widely above the levels of sufficiency for an average demanding crop, so their supply was not required. On the other side, the available phosphorus was low and, hence, it needed to be added with specific fertilizations.

Both composts were used in an open field test. The study compared three fertilization treatments and a control group: mineral fertilization, using diammonium phosphate (18–46); compost from residue recovered from the rearing on a Gainsville diet (GD_{comp}); compost from residue recovered from the rearing on fruit and vegetable waste (FVW_{comp}); and unfertilized control. The amounts of compost and mineral fertilizer were calculated to provide 130 kg/ha of nitrogen as required for processing tomato [48] and provided near the plants just after transplanting.

The experimental design was a 4 × 4 Latin square. Each plot included 5 plants spaced 40 cm apart. A buffer zone of 80 cm was left between the plots. The distance between the rows was 1 m. The total plot area was 1.6 m^2, which corresponds to approx. 31,250 p/ha. To reduce evaporation losses and weed infestation, the area was mulched immediately with wheat straw after transplanting (Figure 1). Plants were drip irrigated with lines placed near the plants and the water supply was scheduled according to the Integrated Production Regulations for processing tomato [48].

Figure 1. Tomato plants at different stages of vegetative growth: 15 (**above**) and 30 (**below**) days after transplant. The picture shows the irrigation system and the mulching on the area.

The same regulations were applied for controlling late blight and insect pests (southern green shield bug and tomato pinworm) affecting the plants during growth. Meteorological data were collected by the Arsial control unit of Monterotondo (RM), location: Grotta Marozza (92 m asl) [49].

Three central plants of each plot were used for field measurements: chlorophyll content (SPAD units) on at least three leaflets of the composed leaf, using a Konica Minolta, Europe SPAD-502 reader; fruit set (ratio between the number of flowers per inflorescence and the number of fruits) of the first two basal inflorescences of each plant

Fruits were collected after 98 days from transplanting when around 80% of the fruits were at the red-ripe stage. The number of marketable, unripened, and damaged fruits per plant was recorded and weighed using a digital balance (Mettler PC8000, Mettler-Toledo S.p.A., Milan, Italy). Ten marketable fruits for each plot were chosen to determine the average length and the polar diameter. The ratio of length/diameter was used for the calculation of the fruit shape index (FSI).

Finally, five fruits per plot identified as marketable were used for the destructive analyses. The fruits were squeezed and the juice was used to determine the pH with the SensION pH25+ portable pH meter (Crison Instruments, Alella, Spain) and the soluble solids (°Brix) with the OPTech digital refractometer (Optical Technology, Hannover, Germany). pH and soluble solids were measured in triplicate.

2.5. Statistical Analysis

All the statistical analyses were performed using PAST software version 4.13 [50]. The data concerning the combination between feeding rate and rearing scale were compared using two-way ANOVA; the remaining data related to substrate and larval biomass characterization, maximum larval weight, waste reduction percentage, and fertilization study were analyzed by one-way ANOVA. Before performing the ANOVA test, data were checked for normality by Shapiro–Wilk test and the Levene's test for the homogeneity of variance. Differences among the means were tested according to the Tukey's honestly significant difference (HSD) test ($p < 0.05$). Differences in composition between two types of compost were analyzed by Student's *t*-test. Principal component analysis (PCA) was performed using the same software to visualize the differences between the fertilization treatments.

3. Results

3.1. Trial A: Feeding Regimes Comparison

This study compared two feeding regimes (HFR and LFR) at a medium (MS) and large scale (LS) and was performed with the substrate FVW:BW, identified as number 1. Although the general ratio among the residues was the same (FVW:BW 80:20 in trial A), each substrate supplied to BSFL had specific composition (Table 3). Significant differences among the values of the substrate nutritional components were found for DM, ash, crude lipid, and carbohydrate content. In all the cases, only one type of substrate differs from the others and, with the exception of crude lipids, all the values were within a narrow range. The crude protein content did not significantly vary among the different substrates.

Table 3. Characterization of the substrate (mean ± SD) used in the experiment at high feeding rate (HFR) and low feeding rate (LFR) with substrate 1 (FVW:BW) in both the medium (MS) and large (LS) scales. Different letters indicate statistically significant differences for Tukey's HSD test ($p < 0.05$).

Parameter	U.M. [1]	Scale	Trial	
			HFR1	LFR1
DM [2]	%	MS	25.4 ± 5.9 a	27.3 ± 3.6 a
		LS	31.7 ± 1.7 a	21.9 ± 1.5 b
Ash	g/100 gDM	MS	4.8 ± 0.2 a	4.9 ± 0.3 a
		LS	5.2 ± 0.2 a	3.8 ± 0.6 b
Crude lipid	g/100 gDM	MS	1.8 ± 0.1 b	1.6 ± 0.3 b
		LS	4.7 ± 0.3 a	2.3 ± 0.6 b
Crude protein	g/100 gDM	MS	13.2 ± 0.3 a	12.8 ± 0.6 a
		LS	13.1 ± 0.5 a	14.6 ± 0.9 a
Carbohydrate	g/100 gDM	MS	80.5 ± 0.5 b	81.2 ± 0.2 a
		LS	78.0 ± 0.9 b	79.3 ± 0.5 b
P:C [3] ratio		MS	1:6	1:6.5
		LS	1:6	1:5.5

[1] U.M., unit of measurement; [2] DM, dry matter; [3] P:C, protein/carbohydrate.

The variability in substrates is due to the variability in FVW itself and in the type of BW, supplies that have not been chosen, coming directly from the supermarket. With regard to BSFL nutritional needs, all the substrates had a low lipid content (range 1.6 ± 0.3–4.7 ± 0.3 g/100 gDM), a medium–low protein content (range 12.8 ± 0.6–14.6 ± 0.9 g/100 gDM), a very high content of carbohydrates (range 78.0 ± 0.8–81.2 ± 0.2 g/100 gDM), and, consequently, a very unbalanced P:C ratio (range 1:5.5–1:6.5).

The addition of BW was effective in obtaining a suitable substrate for larval growth in terms of moisture content (range 68.3 ± 1.7–78.1 ± 1.5 g/100 gDM).

Larval weight increment is shown in Figure 2 and in Table 4. The trend of larval growth for all the trials looks like a part of a sigmoidal curve [51]. The highest MLW was reached at day 13 in MS-HFR1 and was equal to 368.2 ± 22.0 mg/BSFL value that is significantly different from the MLW reached in 14 days in MS-LFR1 (256.8 ± 21.9 mg/BSFL). In the LS trials, BSFL have longer larval lifetime until mature stage. The MLW was reached at day 17 in LS-HFR1 (347.7 ± 12.8 mg/BSFL) and at day 19 in LS-LFR1 (247.3 ± 1.2 mg/BSFL). MLW values at HFR were significantly higher than at LFR1 in both scales. The shift from MS to LS seems to be linear. Regarding the feeding regime, the batch feeding strategy does not allow the quantity of feed provided to a single larva to be exactly modulated. At the end of the experiment, based upon the time necessary to achieve the MLW, we could verify the effective feeding rate intended as g/BSFL/day. The range was 0.066–0.089 mg/BSFL/day for LFR against 0.12–0.15 mg/BSFL/day for HFR (Table 4).

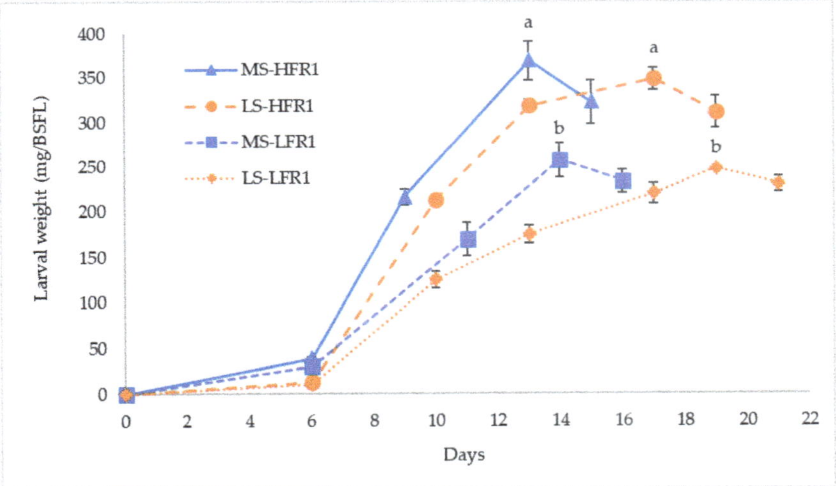

Figure 2. Trend of larval weight (mg/BSFL) during larval lifetime until mature stage (days) comparing medium scale (MS) with large scale (LS) and high feeding regime (HFR) with low feeding regime (LFR) using substrate 1 (FVW:BW). Bars indicate standard deviation. Within MS or LS condition different letters indicate statistically significant differences for Tukey's HSD test ($p < 0.05$).

Regarding waste reduction (Figure 3), statistically significant highest values were obtained in MS-LFR1 (79.3 ± 1.2%) and in LS-HFR1 (77.4 ± 3.0%). In this case, the shift toward higher production scales was opposite and nonlinear; the waste reduction increased significantly passing from MS to LS with the HFR, while decreasing likewise significantly from MS to LS using the LFR.

Table 4. Maximum larval weight (mean ± SD), larval lifetime until mature stage, and effective feeding rate: comparison between high feeding rate (HFR) and low feeding rate (LFR) with substrate 1 (FVW:BW) in both the medium (MS) and large (LS) scale. Different letters beside the maximum larval weight indicate statistically significant differences for Tukey's HSD test ($p < 0.05$).

Parameter	U.M. [1]	Scale	Trial	
			HFR1	**LFR1**
Maximum larval weight	mg/BSFL	MS LS	368.2 ± 22.0 a 347.7 ± 12.2 a	256.8 ± 19.3 b 247.3 ± 1.2 b
Time until mature stage	days	MS LS	13 17	14 19
Effective feeding rate	mg/BSFL/day	MS LS	0.15 0.12	0.089 0.066

[1] U.M., unit of measurement.

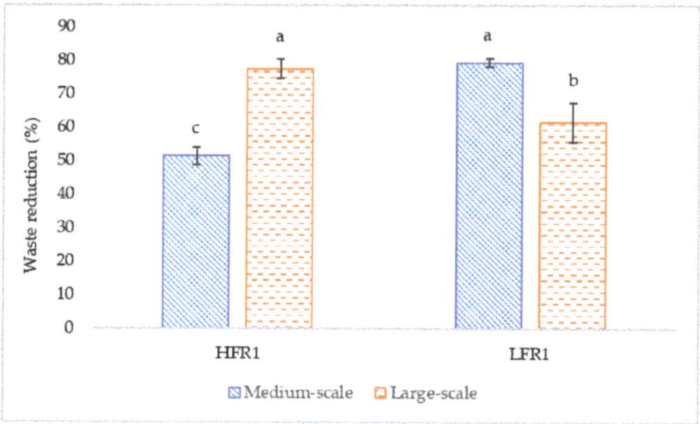

Figure 3. Waste reduction (% on wet waste): comparison between high feeding rate (HFR) and low feeding rate (LFR) with substrate 1 (FVW:BW) in both medium (MS) and large (LS) scale. Bars indicate standard deviation. Different letters indicate statistically significant differences for Tukey's HSD test ($p < 0.05$).

The characterization of BSFL biomass (Table 5) showed statistically significant differences between HFR1 and LFR1 at both scales. In detail, the significant highest values were obtained, both in MS and LS, at HFR1 for the percentage of DM and the crude protein content, while for the percentage of ash and crude lipid content at LFR1. Evident linearity between MS and LS has been detected in HFR1 for DM, ash, lipid, protein, and carbohydrate content per larva. Differently, at the same feeding regime, none of the values related to the other parameters (PrCR, total larval biomass, total lipids, and proteins) showed linearity in the passage from MS to LS. In contrast, at LFR1, biomass and protein production maintains linearity. With regard to PrCR, the highest value was obtained in MS-HFR1 (58.8 ± 5.1%), which was significantly different from LS-HFR1 (37.7 ± 2.3%). At HFR, for the parameters associated to global production (biomass, lipids, and proteins production), the values obtained at MS are significantly higher than those obtained at LS. In both the scales, LFR1 achieves good results in terms of biomass and lipid production comparable or significantly better than HFR1 (Table 5). At LFR, moving from MS to LS did not appear to significantly change biomass and protein production. At the same time, the choice of LFR at LS condition gave a significant increase in lipid content. LFR appeared to have a buffering effect in scaling up from MS to LS. Keeping in mind the objective of scaling up the system at an affordable cost and considering the results with LFR, trial B was set to compare the bioconversion efficiency in MS and LS with other substrates at LFR condition.

Table 5. BSFL biomass characterization and total amount of BSFL biomass and its added-value products (mean ± SD): comparison between HFR1 (high feeding rate) and LFR1 (low feeding rate) with substrate 1 (FVW:BW) in both medium (MS) and large (LS) scale. Different letters indicate statistically significant differences for the Tukey's HSD test ($p < 0.05$).

Parameter	U.M. [1]	Scale	Trial	
			HFR1	LFR1
DM [2]	%	MS	34.5 ± 0.8 a	28.7 ± 0.6 b
		LS	33.8 ± 0.6 a	27.6 ± 0.8 b
Ash	g/100 gDM	MS	4.8 ± 0.1 b	5.7 ± 0.5 a
		LS	4.2 ± 0.3 b	4.9 ± 0.3 a
Crude lipid	g/100 gDM	MS	38.1 ± 0.5 b	39.7 ± 2.9 b
		LS	41.7 ± 1.0 b	45.7 ± 1.0 a
Crude protein	g/100 gDM	MS	31.7 ± 0.5 a	27.4 ± 2.2 b
		LS	29.2 ± 0.4 a	23.7 ± 0.5 c
Carbohydrate	g/100 gDM	MS	25.7 ± 0.5 a	27.1 ± 1.2 a
		LS	24.9 ± 0.4 a	25.8 ± 0.7 a
PrCR [3]	%	MS	58.8 ± 5.1 a	46.5 ± 2.9 b
		LS	37.7 ± 2.3 c	42.4 ± 1.5 bc
Biomass production	g/kgWW [4]	MS	61.7 ± 5.3 a	57.3 ± 2.0 a
		LS	49.5 ± 3.1 b	57.3 ± 1.1 a
Lipid production	g/kgWW	MS	23.5 ± 2.3 ab	22.8 ± 2.4 ab
		LS	20.7 ± 1.4 b	26.0 ± 0.3 a
Protein production	g/kgWW	MS	19.6 ± 1.7 a	15.7 ± 1.0 b
		LS	14.5 ± 0.9 b	13.6 ± 0.5 b

[1] U.M., unit of measurement; [2] DM, dry matter; [3] PrCR, protein conversion ratio; [4] WW, wet waste.

3.2. Trial B: Medium-Scale and Large-Scale Comparison

Trial B aimed to compare MS with LS using all the substrates described in Table 1 (FVW:BW, FVW:BR, FVW:BSG, and FVW:DS) at LFR. Substrate characterization is shown in Table 6.

Table 6. Characterization of the substrates in the experiment with low feeding rate (LFR) in both medium (MS) and large (LS) scale. Numbers 1, 2, 3, and 4 refer to the type of substrate described in Section 2.2. and Table 1 (mean ± SD). Different letters within each parameter indicate statistically significant differences for Tukey's HSD test ($p < 0.05$).

Parameter	U.M. [1]	Scale	Trial			
			LFR1	LFR2	LFR3	LFR4
DM [2]	%	MS	27.3 ± 3.6 a	25.3 ± 0.5 ab	16.9 ± 1.0 c	8.9 ± 0.6 d
		LS	21.9 ± 1.0 b	26.8 ± 0.6 a	15.6 ± 1.1 c	11.5 ± 1.3 d
Ash	g/100 gDM	MS	4.9 ± 0.3 bc	8.4 ± 0.1 a	5.2 ± 0.0 bc	9.2 ± 0.0 a
		LS	3.8 ± 0.6 c	6.2 ± 1.2 b	8.3 ± 0.0 a	9.5 ± 0.0 a
Crude lipid	g/100 gDM	MS	1.6 ± 0.1 d	2.9 ± 0.1 c	4.4 ± 0.1 b	2.8 ± 0.1 c
		LS	2.3 ± 0.2 cd	2.6 ± 0.3 c	6.0 ± 0.4 a	6.3 ± 0.5 a
Crude protein	g/100 gDM	MS	12.4 ± 0.0 d	20.9 ± 1.3 a	17.8 ± 0.5 b	21.3 ± 1.4 a
		LS	14.6 ± 0.9 c	17.5 ± 0.5 b	15.2 ± 0.2 c	14.9 ± 0.2 c
Carbohydrate	g/100 gDM	MS	81.2 ± 1.2 a	67.8 ± 1.3 c	72.5 ± 0.4 b	66.7 ± 1.4 c
		LS	79.3 ± 0.5 a	73.7 ± 0.7 b	70.6 ± 0.2 bc	69.7 ± 2.7 bc
P:C [3] ratio		MS	1:6.5	1: 3	1:4	1:3
		LS	1:5.5	1: 4	1:4.5	1:4.5

[1] U.M., unit of measurement; [2] DM, dry matter; [3] P:C, protein/carbohydrate.

From the observation of the data of Table 6, it appears that the addition of BW and BR (LFR1 and LFR2, respectively) to the FVW was effective in obtaining a substrate with a moisture content ranging from 72.7 ± 3.6% to 78.1 ± 1.2%, suitable for larval growth. In contrast, the addition of BSG (LFR3) and DS (LFR4) absorbed only a small part of the leachate, making a substrate with a moisture content ranging from to 83.2 ± 1.3% to 91.1 ± 0.9%.

Regarding crude lipids, the difference was significant in LFR3 and LFR4, while, for protein content, there was a statistical difference in all substrates,

The carbohydrate content was comparable within each substrate except for LFR2. However, the range between the substrates of the MS and LS tests was very narrow. Moreover, owing to the BW presence in LFR1 the carbohydrate content was statistically lower in substrates 2, 3, and 4. As a consequence, P:C ratio lowered proportionally.

In general, considering the nutritional needs of BSFL, all the substrates had an excess of carbohydrates content. The higher protein content of substrate 2, 3, and 4 compared to substrate 1 contributed to increasing the P:C ratio ranging from 1:3 to 1:4.5.

The differences within the same substrate are due, as already described, to the variability in FVW itself and to the use of different by-products mixed with FVW.

The MLWs in substrates 2, 3, and 4, in both MS and LS trials, were significantly lower than values reached on substrate 1 (Figure 4) and the lowest ones were those of FVW:DS (LFR4). Within each substrate, no statistically significant differences were observed between MS and LS, except in FVW:BR (LFR2). The MLW in the LS test was over 200 mg in both LFR2 (216.0 ± 5.3 mg/BSFL) and LFR3 (206.7 ± 3.2 mg/BSFL). Except LFR2, the larval lifetime until the mature stage was longer in LS than MS; for the first, it ranged from 16 to 19 days, while, at MS, the range was 10–14 days. The feeding rate resulted in being effectively at a low level (ranged from 0.066 in LS-LFR1 to 0.125 mg/BSFL/day in MS-LFR4), which was the condition we wanted to test. Waste reduction at MS condition (Figure 5) showed a decreasing trend according to the order LFR1 > LFR2 > LFR4 > LFR3. The same pattern was partially confirmed for the LS where LFR1 and LFR2 had a comparable waste reduction, higher than the waste reduction observed for LFR3 and LFR4. Waste reduction in MS was significantly higher (LFR1), lower (LFR3), or comparable (LFR2 and LFR3) compared to the LS condition.

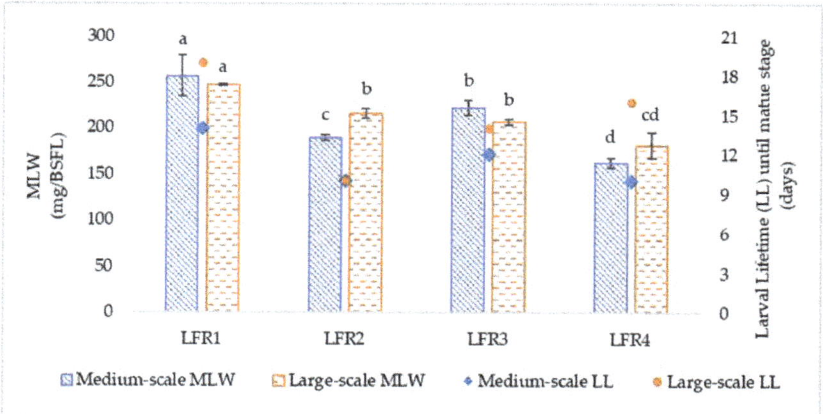

Figure 4. Maximum larval weight (MLW) (mg/BSFL) and larval lifetime (LL) until mature stage (days) in the experiment with low feeding rate (LFR) in both medium (MS) and large (LS) scale. Numbers 1, 2, 3, and 4 refer to the type of substrate described in Section 2.2 and Table 1. Bars indicate standard deviation. Different letters indicate statistically significant differences for Tukey's HSD test ($p < 0.05$).

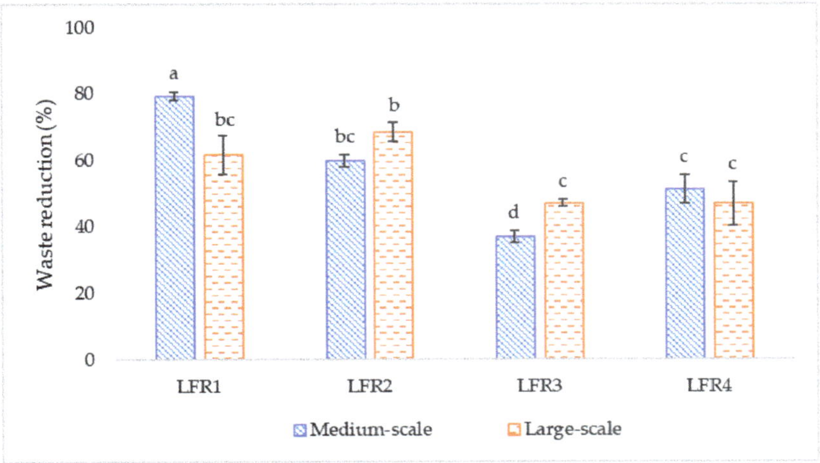

Figure 5. Waste reduction (% on wet waste) in the experiment with low feeding rate (LFR) in both medium (MS) and large (LS) scale. Numbers 1, 2, 3, and 4 refer to the type of substrate described in Section 2.2 and Table 1. Bars indicate standard deviation. Different letters indicate statistically significant differences for Tukey's HSD test ($p < 0.05$).

LFR1 and LFR2 showed the highest values of DM, independently of the rearing scale (Table 7). LFR4 had the lowest DM. Interestingly, LFR1 was the substrate allowing the highest accumulation of crude lipid in the larvae both in MS (39.7 ± 2.8%) and LS (45.7 ± 1.0%). The difference with the other substrates was very impressive, ranging from 1.5 (MS-LFR3) to almost fourfold (MS-LFR4) compared to the highest value (LS-LFR1). On the other side, the content of crude protein and carbohydrate in LFR1 were lower than the remaining three substrates. Specifically, the percentage of crude protein was abundantly above 30% in LFR2, LFR3, and LFR4, while, in LFR1, the BSFL crude protein was just below this threshold in MS (27.4 ± 2.2%) and even significantly lower in LS (23.7 ± 0.5%). LFR2 and LFR4 were instead the substrates allowing for the highest accumulation of carbohydrates in BSFL. Except LFR1, where MS and LS were comparable, the highest values of the PrCR value were obtained in the LS condition, with a difference statistically significant compared to the MS.

Table 7. BSFL biomass characterization: comparison between medium- and large-scale trials with the FVW-based substrates (mean ± SD) in the experiment with low feeding rate (LFR) in both medium (MS) and large (LS) scale. Numbers 1, 2, 3, and 4 refer to the type of substrate described in Section 2.2 and Table 1. Different letters indicate statistically significant differences for the Tukey HSD test ($p < 0.05$).

Parameter	U.M. [1]	Scale	Trial			
			LFR1	LFR2	LFR3	LFR4
DM [2]	%	MS LS	28.7 ± 0.6 a 27.6 ± 0.8 a	24.8 ± 0.2 b 27.3 ± 0.3 a	25.0 ± 0.2 b 24.1 ± 0.9 b	16.5 ± 0.2 d 20.8 ± 0.7 c
Ash	g/100 gDM	MS LS	5.7 ± 0.5 d 4.9 ± 0.3 d	8.9 ± 0.2 c 5.5 ± 0.5 d	7.5 ± 0.3 c 8.6 ± 0.5 c	17.1 ± 1.0 a 10.5 ± 0.3 b
Crude lipid	g/100 gDM	MS LS	39.7 ± 2.9 b 45.7 ± 1.0 a	21.1 ± 1.8 e 27.9 ± 1.3 cd	30.6 ± 1.2 c 26.8 ± 1.1 cd	12.8 ± 0.9 f 24.8 ± 0.3 de
Crude protein	g/100 gDM	MS LS	27.4 ± 2.2 c 23.7 ± 0.5 d	37.1 ± 1.3 a 35.8 ± 0.4 ab	34.0 ± 0.2 ab 34.5 ± 0.8 ab	34.4 ± 0.7 ab 32.9 ± 0.8 b
Carbohydrate	g/100 gDM	MS LS	27.1 ± 1.2 d 25.8 ± 0.7 d	33.0 ± 0.3 b 30.8 ± 1.4 bc	27.9 ± 1.6 cd 30.1 ± 0.3 c	35.8 ± 0.9 a 31.8 ± 0.3 bc

Table 7. *Cont.*

Parameter	U.M. [1]	Scale	Trial			
			LFR1	LFR2	LFR3	LFR4
PrCR [3]	%	MS	46.5 ± 2.9 bc	27.6 ± 0.8 f	46.3 ± 5.1 ac	36.5 ± 1.6 de
		LS	42.4 ± 1.5 cd	34.9 ± 1.2 e	53.0 ± 2.4 ab	53.8 ± 1.8 a
Biomass production	g/kgWW [4]	MS	57.3 ± 2.0 a	39.4 ± 2.5 c	40.9 ± 4.3 bc	20.2 ± 1.2 e
		LS	57.3 ± 1.1 a	46.1 ± 1.4 b	36.5 ± 0.8 c	27.8 ± 0.9 d
Lipid production	g/kgWW	MS	22.8 ± 2.4 a	8.3 ± 1.2 c	12.6 ± 1.7 b	2.6 ± 0.3 d
		LS	26.0 ± 0.3 a	12.8 ± 1.0 b	9.78 ± 0.2 bc	6.9 ± 0.3 c
Protein production	g/kgWW	MS	15.7 ± 1.0 ab	14.6 ± 0.4 bc	13.9 ± 1.5 bc	6.9 ± 0.3 e
		LS	13.6 ± 0.5 c	16.5 ± 0.3 a	12.6 ± 0.6 c	9.2 ± 0.3 d

[1] U.M., unit of measurement; [2] DM, dry matter; [3] PrCR, protein conversion ratio; [4] WW, wet waste.

When the results were expressed as grams per kilogram of wet waste, the data emphasized the positive effect of the FVW:BW in LFR1 compared to the other substrates (Table 7). The values of biomass and lipid production were statistically different from those shown by the other substrates, while, for protein production, the highest values were measured for MS-LFR1 (15.7 ± 1.0 g/kgWW) and LS-LFR2 (16.5 ± 0.3 g/kgWW).

3.3. Field Trial

The weather showed a typical trend of the area (Figure 6). Substantial rainfall occurred until early June, gradually reducing in the summer months until harvest. Transplanting at the end of May made it possible to escape the downy mildew infestations that have damaged early tomato transplants in Italy. Gradually increasing temperatures through July mirrored the trends observed for central Italy by requiring proper water supply at different stages of plant development.

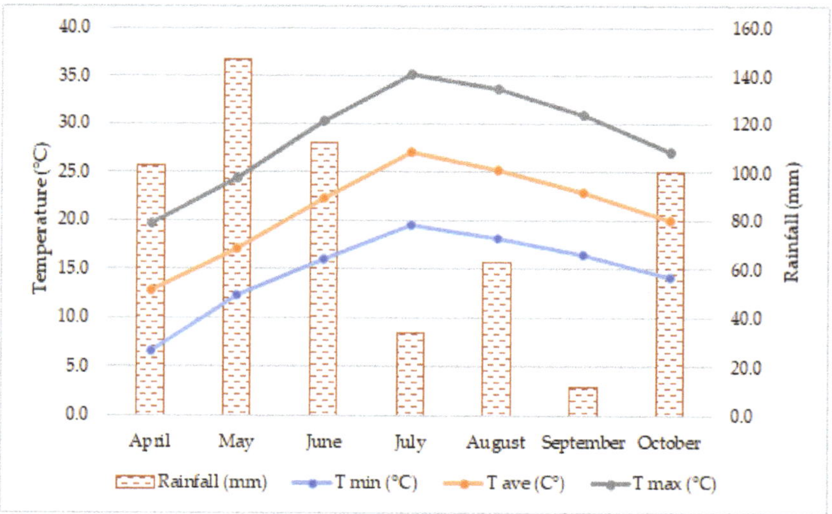

Figure 6. Data on minimum, average, maximum temperatures, and monthly rainfall of the period April–November 2023 were collected by the Arsial control unit of Monterotondo (RM), location: Grotta Marozza (92 m asl) [49].

The two types of compost were quite different (Table 8). The final moisture and pH of the FVW_{comp} were significantly higher than those of GD_{comp}. Although the carbon (C) and hydrogen (H) content were similar in both types, the nitrogen (N) content was significantly lower in FVW_{comp}. As a result, the C/N ratio in FVW_{comp} was significantly higher than that of GD_{comp}. It should be noted that, in the straw added during the composting process, the average C content was 47.4 ± 1.6% and that of N 0.6 ± 0.1%, with a C:N ratio of 76.8 ± 8.6. The data measured in the mature compost indicated an effect of composting in obtaining a C:N ratio acceptable for agronomic use. Except for iron (Fe), the content of some macro- and microelements was significantly higher in FVW_{comp}.

Table 8. Characterization (mean ± SD; n = 3) of GD and FVW compost.

Value	U.M. [a]	GD_{comp} [b]	FVW_{comp}	t-Value	p [c]
Moisture	%	16.9 ± 4.1	48.2 ± 5.7	7.73	**
pH		6.9 ± 0.2	8.3 ± 0.4	6.17	**
C	%	34.9 ± 4.5	39.8 ± 1.0	1.86	ns
H	%	7.1 ± 1.7	7.6 ± 1.6	0.36	ns
N	%	2.8 ± 0.3	1.9 ± 0.3	3.70	*
C/N		12.4 ± 0.4	21.7 ± 2.6	6.22	**
Na	g/kg	0.9 ± 0.2	1.5 ± 0.3	2.82	*
Mg	g/kg	6.6 ± 0.5	8.3 ± 0.4	4.54	*
K	g/kg	1.4 ± 0.3	2.1 ± 0.7	1.47	ns
Ca	g/kg	5.4 ± 0.5	6.2 ± 1.3	0.95	ns
Fe	g/kg	0.8 ± 0.1	0.5 ± 0.0	3.39	*
Ni	mg/kg	14.5 ± 3.0	17.7 ± 2.7	1.41	ns
Cu	mg/kg	14.9 ± 0.9	20.5 ± 2.8	3.37	*
Zn	mg/kg	66.2 ± 3.7	87.0 ± 6.2	5.01	**
As	mg/kg	4.3 ± 0.8	3.2 ± 1.3	1.26	ns
Cd	mg/kg	0.1 ± 0.0	0.2 ± 0.0	9.96	**
Pb	mg/kg	9.0 ± 1.6	11.1 ± 3.0	1.12	ns

[a] U.M., unit of measurement. [b] GD = Gainsville diet, FVW = fruit and vegetable waste. [c] ns = not significant, * significant at $p \leq 0.05$, ** significant at $p \leq 0.01$ after Student's *t*-test.

In general, no significant differences were observed among the treatments. Compost seems to have a slightly depressive effect on plant development (Table 9) but the chlorophyll content of unfertilized plants was lower than in the other three treatments. The same trend was observed for the fruit set in the second truss, while, in the first truss, there was a positive effect of both chemical fertilization and FVW compost.

Table 9. Main characteristics of tomato plants (mean ± SD).

Treatment [a]	Plant Height (cm)	Chlorophyll Content (SPAD Unit)	Fruit Set	
			1st Truss	2nd Truss
Unfertilized	70.7 ± 5.1	44.7 ± 2.1	63.7 ± 22.1	60.1 ± 10.9
Fertilized	70.7 ± 3.2	47.3 ± 4.9	74.2 ± 4.7	66.1 ± 19.3
GD_{comp}	67.9 ± 6.1	46.5 ± 3.6	66.2 ± 14.8	67.2 ± 4.8
FVW_{comp}	66.7 ± 5.6	46.8 ± 3.0	71 ± 7.1	67.6 ± 8.1

[a] GD_{comp} = compost from Gainsville diet, FVW_{comp} = compost from fruit and vegetable waste.

From a production point of view, there were also no significant differences between treatments (Figure 7), although a definite trend can be appreciated. In terms of the production of marketable fruits, the average fruit weight ranged from 70 (unfertilized control) to 74 g (GD_{comp}). However, the most noticeable differences were in the number of fruits produced per plant, which ranged from 24 (fertilized) to 15 (GD_{comp}). The GD_{comp} figure was confirmed by the high incidence (by weight) of unripened and damaged fruits, which reduced the percentage of marketable fruits to 76%. The final figure indicates decreasing productivity according to the order Fertilized > FVW_{comp} > Unfertilized > GD_{comp}.

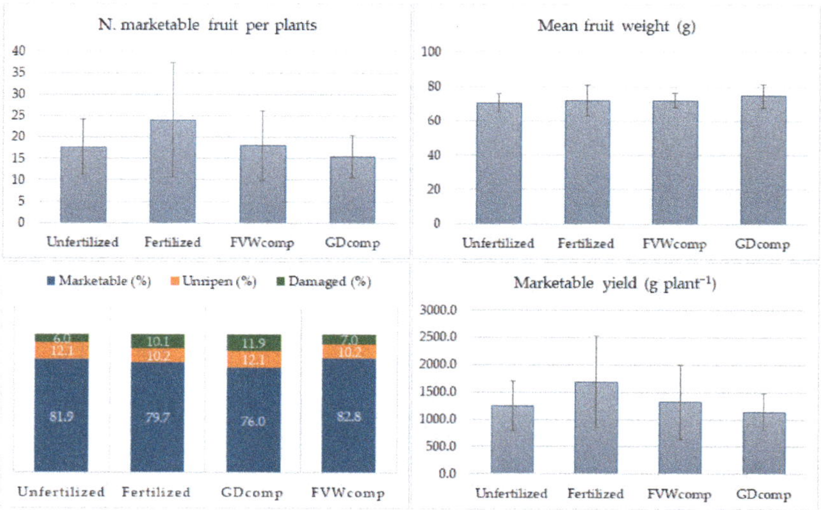

Figure 7. Productive traits observed for the studied treatment. Where present, bars represent the standard deviation. GD_{comp} = compost from Gainsville diet, FVW_{comp} = compost from fruit and vegetable waste.

The characteristics of the fruit did not show significant differences either (Table 10). The fruit length of the unfertilized plants was slightly lower than the other treatments, while the fruits treated with FVW_{comp} had the lowest diameter. The pH value was around 4.6 for all treatments, while there was a slight difference in soluble solids between the plants treated with compost and the fertilized or unfertilized plants.

Table 10. Mean (±SD) of marketable fruit weight, pH, and °Brix.

Treatment [a]	Fruit Traits			pH	Soluble Solids
	Lenght (cm)	Diameter (cm)	FSI [b]		°Brix
Unfertilized	6.6 ± 0.2	4.8 ± 0.2	1.4 ± 0.0	4.6 ± 0.2	5.2 ± 0.7
Fertilized	6.8 ± 0.2	5.0 ± 0.3	1.4 ± 0.0	4.7 ± 0.2	5.2 ± 0.6
GD_{comp}	6.8 ± 0.2	5.1 ± 0.2	1.3 ± 0.0	4.6 ± 0.3	5.0 ± 0.3
FVW_{comp}	6.6 ± 0.3	4.8 ± 0.3	1.4 ± 0.0	4.6 ± 0.2	5.0 ± 0.7

[a] GD_{comp} = compost from Gainsville diet, FVW_{comp} = compost from fruit and vegetable waste. [b] FSI = fruit shape index.

The analysis provided by PCA (Figure 8) confirmed how the different treatments do not clearly differ from each other. The main productive traits (such as number and weight of marketable fruits) were associated with the fertilized treatment or FVW_{comp} application. On the other hand, fruit characteristics and the number of immature fruits were shifted more toward GD_{comp}. Overall, the two main components explain 61.9% of the total variability.

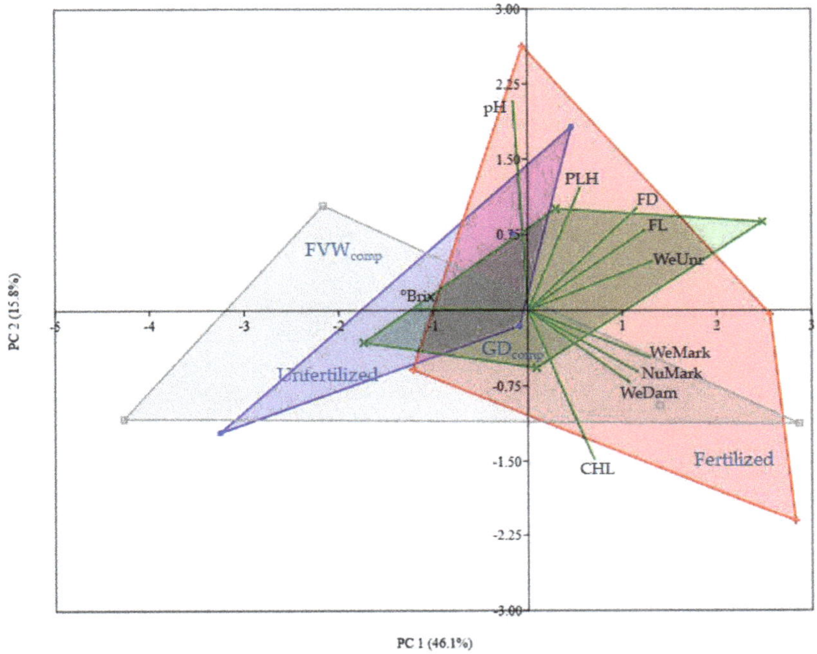

Figure 8. Biplot showing the PCA results of treatment separation based on the observed traits. °Brix (soluble solids); CHL, chlorophyll content; FD, fruit diameter; FL, fruit length; NuMark, number of marketable fruits per plant; PLH, plant height; WeMark, weight of marketable fruits per plant; WeUnr, weight of unripened fruits per plant; WeDam, weight of damaged fruits per plant. Polygon colors: purple, Unfertilized; pink, Fertilized; green, GD$_{comp}$, compost from Gainsville diet; grey, FVW$_{comp}$, compost from fruit and vegetable waste.

4. Discussion

This aimed to analyze the influence of rearing conditions (feeding regimes and rearing scales) on bioconversion efficiency parameters of BSFL in view of system scale-up, comprising the whole supply chain till the field evaluation of the composted frass. Fruit- and vegetable-based diets represent the largest amount of useable organic waste from large-scale retail trade and from local markets, together with a wide range of by-products from the agrifood industry.

FVW supply from the supermarket was heterogeneous. The heterogeneity was not related to the seasonality because of the continuous presence of greenhouse-grown products provided off-season. A wide range of fruits and vegetables (domestic and exotic) were provided by the supermarket and used in the various tests in the different compositions and proportions as they arrived. Similarly, the supply of BW was very variable in quantity (sometimes huge, sometimes few, since it was as a priority targeted to ethical food banking) and quality (pizza and sweets contain fats and carbohydrates). For this reason and to respect a real scenario of a continuous supply by several supermarkets to a CORS-based bioconversion plant mediated by BSFL, the qualitative and quantitative compositions of each supply were not considered as a variable.

The addition of bakery waste (BW) or bran (BR) was effective in lowering the initial FVW moisture content, absorbing a part of the excess leachate, and facilitating the bioconversion process. In contrast, the addition of brewer's spent grain (BSG) was not sufficient to lower the moisture content to the needs of BSFL. A combination of the three types of waste (FVW, BW, and BSG) could represent an improvement in the composition of the substrate and thus the efficiency of the bioconversion process. In addition to the moisture content, the feeding rate (amount of substrate/larva/day) is an important factor that influences the bioconversion process. Diener et al. [14] evaluated the effects of different feeding rates on waste reduction and final biomass production, demonstrating that the optimal value was 0.1 g/larva/day if a balanced feed from the nutritional point of view was ensured. Parra Paz et al. [21] estimated as the ideal condition of BSFL feeding rate a range between 0.095 and 0.163 mg/larva/day. The present study demonstrated that, with the combination FVW:BW, valuable yields of larval biomass and added-value products can be obtained at a low feeding rate (0.066–0.089 g/larva/day), no matter the production scale. Higher yields are attainable (as observed also in our work) with high feeding rate but, when applied to a firm scale, it means higher operating cost. Thus, a trade-off between valuable production and wise economic management must be considered.

BSFL can be reared on a wide range of organic substrates [34] but the bioconversion process is influenced by feedstock characteristics [10]. It was shown that dietary protein and nonstructural carbohydrate contents are primary determinants of bioconversion efficiency [17,22]. Nevertheless, high dietary protein levels do not necessarily lead to higher larval performance [12] since the amount, quality, and ratio of protein, carbohydrates, and lipid should be considered as well [52]. Compared to other studies [53] and considering BSFL nutritional needs [22], the substrates used in this work were poor in protein and lipid content, with a high percentage of carbohydrates. This led to a P:C ratio higher than the optimal rates of 1:2–1:3 [11,54]. Nevertheless, mature larval weights did not differ from those observed by other authors using FVW as the main substrate for BSFL rearing [20,55,56]. At a low feeding rate, the maximum larval weight (MLW) was around 250 mg/BSFL for FVW:BW and remained around 200 mg/BSFL with FVW:BSG and FVW:BR mixtures. The best results in terms of MLW were achieved by using a substrate with a P:C ratio of 1:5.5–1:6.5, which means an unbalanced diet, with a high percentage of carbohydrates. Such data suggest that BSFL are able to exploit poor nutritional diets to reach a body mass comparable to supplying a richer diet, such as kitchen waste with a high level of protein and fat [57].

Shorter development time is especially important, as it leads to less production of greenhouse gases such as CO_2 and NH_3 during the waste bioconversion process by BSFL [58]. In this work, the larval lifetime until the mature stage (i.e., at highest larval weight) was influenced by the scale of the trials, confirming how the transition from laboratory to mass rearing of insects could affect this parameter [59,60]. In the MS trial, BSFL had a development time not exceeding 14 days, while, in LS condition, the lifetime lasted up to 19 days. The weight decrease following the MLW is typical of the prepupae, the last larval stage before pupation. In this phase, the larvae empty their digestive tube, do not feed anymore, and move away from the food source towards a place to pupate [10]. Similarly, Caruso et al. [61] obtained mature larvae under mass production conditions in a time (14 days) longer than under laboratory conditions (9 days). Arabzadeh et al. [62] evaluated that the required time to attain 40% prepupae was 15 days. In their study, they suggested that the faster development time could be due to higher levels of proteins in the diet rather than the presence of BSG and bread in addition to FVW. Other studies reported that BSFL fed with vegetable-based diets completed their larval cycle in 36–52 days [34,55]. Meneguz et al. [63] found a significant difference in the development time comparing BSG diet (8 days) with FV diets (20–22 days), where BSG has a considerably higher protein content. However, protein level was not the key factor in our study, because, despite having a low protein content compared to the carbohydrates, the percentage of larval crude protein was in agreement with those of Arabzadeh et al. [62].

Although the addition of BW and BR did not show linearity between MS and LS trials, probably due to the better conditions in terms of moisture content, the percentage of waste reduction was the highest, ranging from 59.8 to 79.3%. These results are in line with Arabzadeh et al. [62] and Candian et al. [20], who showed that, with the addition of bread to FVW, the substrate reduction reached values of 66.8 and 76.9%, respectively.

In the current study, BSFL reared on FVW with added BW, in both the scales, had the highest crude lipid content. The role of carbohydrates-to-lipids conversion has been well documented [53,64]. Again, the results are nonlinear between MS and LS trials; only in LS does the lipid content of BSFL appear to be affected by the carbohydrate content of the substrates, with the lowest values in BSFL fed with FVW:BSG or FVW:DS. These results, ranging from 24.8 ± 0.3 to 45.7 ± 1.0%, are similar to values found in other studies for FVW-fed larvae [62,65]. Larval crude protein content, ranging from 23.7 ± 0.5 to 37.1 ± 1.3, with the lowest values in FVW:BW, appeared to mirror the protein content of the substrate. This result is in contrast with the findings of Spranghers et al. [53] and Tschirner & Simon [66], who reported that crude protein was higher in larvae fed on a diet with the lowest crude protein content. Nevertheless, the percentages of BSFL crude protein are in agreement with other authors [9,63].

A significantly lower percentage of crude protein content of substrate than BSFL confirmed the efficient capacity of larvae to convert dietary proteins into larval protein biomass [22,34]. The trend of PrCR values was not perfectly comparable between the medium and large scale. In MS trials, the highest values were obtained for FVW:BW and FVW:BSG, while, in LS trials, for FVW:BSG and FVW:DS. In addition, in LS condition, the PrCR showed higher values than in MS. In any case, the PrCR values were in line with the results of Lalander et al. [34] and Arabzadeh et al. [62], who tested FVW diets for BSFL feeding. Lower efficiency in PrCR values in the MS trial and, in general, the disagreement between the scales could be explained by the influence of the scale. At the laboratory scale, with small amounts of larvae, BSFL performance could be more affected by rearing conditions (substrate, temperature, and relative humidity), with a reduction in bioconversion efficiency [65]. These results show how relevant the scale factor is for BSFL rearing and, as for key performance variables (e.g., waste reduction, lipid content, and PrCR), the transition from a benchtop to a large/industrial scale may not be linear [30,67]. In the setting up of the experiments, it was not possible to maintain the same proportion between the quantity of feed provided to BSFL and the size of the container. In LS trials, the fresh food column (about 12–13 cm) was reduced to 4–5 cm during the bioconversion process because of larval activity and water evaporation. Considering the number of starting larvae and the surface area of the container, the final density was about 3.5 BSFL per cm^2. In the MS trials, the larvae were more affected by the weight of the food column weighing on them. In a preliminary test we observed that the same rearing conditions produced much higher amounts of leachate than in LS. Under these conditions, the larvae were unable to co-operate in bioconversion easily because of the lack of available oxygen. Therefore, the food column was halved to about 6–7 cm, halving the number of larvae and the total amount of food. This way, we had about 1.7 BSFL per cm^2, which is still within the optimal range of larval density obtained by Parra Paz et al. [21] between 1 and 5 BSFL/cm^2. Therefore, the conditions for the best performance of the bioconversion process and the parameters to be set (whether they are expressed as feeding regime, larval density, or feeding rate) cannot necessarily be the same on a small, medium, and large scale, thus causing the nonlinearity of the scaling up. Results on BSFL rearing may provide the basis to compare findings from previous small-scale studies. Moreover, they are a paradigm to help in optimizing the mass-rearing conditions of BSFL fed with FVW and agro-industrial by-products.

The most obvious finding from the results of the field trial is the lack of significance between different treatments. There is a trend that indicates a decreasing productivity according to the order Fertilised > FVW_{comp} > Unfertilized > GD_{comp}. Nevertheless, based on the data collected, it is not possible to state if this is due to an intrinsic difference among the fertilization strategies applied. A possible cause may be associated with the high experimental variability responsible for masking the real differences. However, the magnitude of the observed differences does not appear particularly consistent, and it can therefore be an indication of a real equivalence of effects, in particular, between chemical fertilization and compost administration. Although the tested compost was not able to completely equate the results obtained with the synthetic fertilizer, the yields were comparable. This observation is supported by the work of Anyega et al. [26], who, by comparing composted BSF frass fertilizer (BSFFF), composted brewer's spent grain, commercial organic fertilizer (Evergrow), mineral fertilizer, and some combinations, observed a substantial equivalence between BSFFF and chemical fertilization in terms of tomato yield in both greenhouse and in open field conditions. The two treatments were only overcome by the combination of BSFFF with synthetic fertilizer. The reader must be aware that only the repeated medium- and long-term administration of compost at agronomic or organic doses triggers improvement trajectories affecting both the physical and hydraulic properties of the soil [68] and the bulk soil chemical and biological parameters, thus contributing to decreasing the amount of conventional fertilizer [69].

The results obtained using FVW_{comp} were slightly higher than with GD_{comp}. This could be explained by a better composition in terms of micro- and macro-nutrients, which favored a higher production of marketable fruits. Based on the C/N ratio value, GD_{comp} appeared more mature and, therefore, more stable than FVW_{comp}. From a nutritional point of view, this means a slower nutrient release than FVW_{comp}. Conversely, FVW_{comp}, still not fully mature, could promote a faster release and, therefore, a prompt and greater availability of nutrients. The fact that the production of plants fertilized with GD_{comp} included the highest proportion of unripened fruits would confirm the effect.

5. Conclusions

The primary objective of project Hermes was to create an operating model to valorize the waste of a large supermarket and by-products of the agro- or food industry within the territory of the Lazio Region, carrying out experiments at a large-scale level. The heterogeneity of fruit and vegetable waste coming from a large distribution did not appear a limiting factor about the possibility to standardize the production in an insect farm as well as to organize the bioconversion process at an industrial scale. The study demonstrates that the addition of co-substrate such as bakery waste or bran to fruits and vegetable waste lowers the final moisture content and improves larval growth. At a large scale, an increase in the ratios of larvae to feed, even if it resulted in inducing individual lower larval weight, produced high larval biomass as well as substantial total production of crude proteins and lipids. The transition from medium scale to large scale in this study resulted in being partially linear. So, it is possible to expect that, by modifying the feeding regime (and we noticed that there is room for a reduction) and by studying the proportions between the basic substrate (FVW) and the co-substrates, we can identify a model for the development of an insect farm based on the waste supplies of one or more supermarket chains.

The fertilizer trials found no significant difference between fertilized and unfertilized plants regardless of the type of fertilizer used. However, the data from the field trial provided indications about the possibility of using the compost obtained from the residues of BSF rearing as a soil conditioner even in partial replacement of the amount of synthetic fertilizer. Such a result increases the degree of integration of the bioconversion system within a circular economy context.

Author Contributions: Conceptualization, E.D.S., F.L., M.D.M., E.S. and S.A.; data curation, E.D.S., A.d.I., M.C., C.B. and E.S.; formal analysis, E.D.S. and E.S.; funding acquisition, A.A. and E.S.; investigation, E.D.S., A.d.I., F.L., M.D.M., F.P., A.D.G., M.C., F.G., C.B., E.S. and S.A.; methodology, E.D.S., A.d.I., F.L., M.D.M., C.B., E.S. and S.A.; project administration, A.A., E.S. and S.A.; resources, E.D.S., F.L., M.D.M., F.P., F.G., C.B., A.A., E.S. and S.A.; supervision, E.S. and S.A.; validation, E.D.S., A.d.I., E.S. and S.A.; visualization, E.D.S., E.S. and S.A.; writing—original draft preparation, E.D.S., E.S. and S.A.; writing—review and editing, E.D.S., E.S. and S.A. All authors have read and agreed to the published version of the manuscript.

Funding: This research was funded within POR FESR Lazio 2014–2020 (Det. Reg. n. G09493 140721, 22/07/2021), Project HERMES—"*Hermetia illucens* per il recupero e la valorizzazione di biomassa residuale: realizzazione di un modulo prototipale ecosostenibile".

Institutional Review Board Statement: Not applicable.

Data Availability Statement: All relevant data are contained within the article.

Acknowledgments: We want to thank Unicoop Tirreno supermarket located in Rome, Largo Nino Franchellucci, and the Ritual Lab brewery in Formello (Rome) for sensitivity towards the issue and for ensuring the continuous supply of waste for the duration of the tests. We also thank the colleagues of Bioproducts and Bioprocesses Laboratory of the Enea Casaccia Centre for hosting us in their facilities and allowing us to make the chemical characterization of the waste and larval biomass. A special thank to Massimo Cristofaro, research leader of the non-profit Foundation Biotechnology and Biological Control Agency in Rome (BBCAonlus), for reviewing the manuscript.

Conflicts of Interest: The authors declare no conflicts of interest.

References

1. Abd El-Hack, M.E.; Shafi, M.E.; Alghamdi, W.Y.; Abdelnour, S.A.; Shehata, A.M.; Noreldin, A.E.; Ashour, E.A.; Swelum, A.A.; Al-sagan, A.A.; Alkhateeb, M.; et al. Black Soldier Fly (*Hermetia illucens*) Meal as a Promising Feed Ingredient for Poultry: A Comprehensive Review. *Agriculture* **2020**, *10*, 339. [CrossRef]
2. Food and Agriculture Organization of the United Nations (FAO). The Future of Food and Agriculture: Trends and Challenges. Available online: http://www.fao.org/3/a-i6583e.pdf (accessed on 8 January 2018).
3. Bernstad Saraiva Schott, A.; Andersson, T. Food Waste Minimization from a Life-Cycle Perspective. *J. Environ. Manag.* **2015**, *147*, 219–226. [CrossRef] [PubMed]
4. Ramamoorthy, K.; Dhanraj, R.; Vijayakumar, N.; Ma, Y.; Al Obaid, S.; Narayanan, M. Vegetable and Fruit Wastes: Valuable Source for Organic Fertilizer for Effective Growth of Short-Term Crops: *Solanum lycopersicum* and *Capsicum annuum*. *Environ. Res.* **2024**, *251*, 118727. [CrossRef] [PubMed]
5. Manzano-Agugliaro, F.; Sanchez-Muros, M.J.; Barroso, F.G.; Martínez-Sánchez, A.; Rojo, S.; Pérez-Bañón, C. Insects for Biodiesel Production. *Renew. Sustain. Energy Rev.* **2012**, *16*, 3744–3753. [CrossRef]
6. De Souza-Vilela, J.; Andrew, N.R.; Ruhnke, I. Insect Protein in Animal Nutrition. *Anim. Prod. Sci.* **2019**, *59*, 2029. [CrossRef]
7. Kaya, C.; Generalovic, T.N.; Ståhls, G.; Hauser, M.; Samayoa, A.C.; Nunes-Silva, C.G.; Roxburgh, H.; Wohlfahrt, J.; Ewusie, E.A.; Kenis, M.; et al. Global Population Genetic Structure and Demographic Trajectories of the Black Soldier Fly, *Hermetia illucens*. *BMC Biol.* **2021**, *19*, 94. [CrossRef] [PubMed]
8. Singh, A.; Kumari, K. An Inclusive Approach for Organic Waste Treatment and Valorisation Using Black Soldier Fly Larvae: A Review. *J. Environ. Manag.* **2019**, *251*, 109569. [CrossRef] [PubMed]
9. Giannetto, A.; Oliva, S.; Riolo, K.; Savastano, D.; Parrino, V.; Cappello, T.; Maisano, M.; Fasulo, S.; Mauceri, A. Waste Valorization via *Hermetia illucens* to Produce Protein-Rich Biomass for Feed: Insight into the Critical Nutrient Taurine. *Animals* **2020**, *10*, 1710. [CrossRef] [PubMed]
10. Surendra, K.C.; Tomberlin, J.K.; Van Huis, A.; Cammack, J.A.; Heckmann, L.-H.L.; Khanal, S.K. Rethinking Organic Wastes Bioconversion: Evaluating the Potential of the Black Soldier Fly (*Hermetia illucens* (L.)) (Diptera: Stratiomyidae) (BSF). *Waste Manag.* **2020**, *117*, 58–80. [CrossRef]
11. Barragan-Fonseca, K.B.; Gort, G.; Dicke, M.; Van Loon, J.J.A. Effects of Dietary Protein and Carbohydrate on Life-history Traits and Body Protein and Fat Contents of the Black Soldier Fly *Hermetia illucens*. *Physiol. Entomol.* **2019**, *44*, 148–159. [CrossRef]
12. Gold, M.; Binggeli, M.; Kurt, F.; De Wouters, T.; Reichlin, M.; Zurbrügg, C.; Mathys, A.; Kreuzer, M. Novel Experimental Methods for the Investigation of *Hermetia illucens* (Diptera: Stratiomyidae) Larvae. *J. Insect Sci.* **2020**, *20*, 21. [CrossRef] [PubMed]
13. Laganaro, M.; Bahrndorff, S.; Eriksen, N.T. Growth and Metabolic Performance of Black Soldier Fly Larvae Grown on Low and High-Quality Substrates. *Waste Manag.* **2021**, *121*, 198–205. [CrossRef]
14. Diener, S.; Zurbrügg, C.; Tockner, K. Conversion of Organic Material by Black Soldier Fly Larvae: Establishing Optimal Feeding Rates. *Waste Manag. Res. J. Sustain. Circ. Econ.* **2009**, *27*, 603–610. [CrossRef]
15. Cheng, J.Y.K.; Chiu, S.L.H.; Lo, I.M.C. Effects of Moisture Content of Food Waste on Residue Separation, Larval Growth and Larval Survival in Black Soldier Fly Bioconversion. *Waste Manag.* **2017**, *67*, 315–323. [CrossRef] [PubMed]
16. Bekker, N.S.; Heidelbach, S.; Vestergaard, S.Z.; Nielsen, M.E.; Riisgaard-Jensen, M.; Zeuner, E.J.; Bahrndorff, S.; Eriksen, N.T. Impact of Substrate Moisture Content on Growth and Metabolic Performance of Black Soldier Fly Larvae. *Waste Manag.* **2021**, *127*, 73–79. [CrossRef]
17. Cammack, J.; Tomberlin, J. The Impact of Diet Protein and Carbohydrate on Select Life-History Traits of The Black Soldier Fly *Hermetia illucens* (L.) (Diptera: Stratiomyidae). *Insects* **2017**, *8*, 56. [CrossRef]
18. Lalander, C.; Ermolaev, E.; Wiklicky, V.; Vinnerås, B. Process Efficiency and Ventilation Requirement in Black Soldier Fly Larvae Composting of Substrates with High Water Content. *Sci. Total Environ.* **2020**, *729*, 138968. [CrossRef]
19. Khaekratoke, K.; Laksanawimol, P.; Thancharoen, A. Use of Fermented Spent Coffee Grounds as a Substrate Supplement for Rearing Black Soldier Fly Larvae, *Hermetia Illucens* (L), (Diptera: Stratiomyidae). *PeerJ* **2022**, *10*, e14340. [CrossRef]
20. Candian, V.; Meneguz, M.; Tedeschi, R. Immune Responses of the Black Soldier Fly *Hermetia illucens* (L.) (Diptera: Stratiomyidae) Reared on Catering Waste. *Life* **2023**, *13*, 213. [CrossRef] [PubMed]

21. Parra Paz, A.S.; Carrejo, N.S.; Gómez Rodríguez, C.H. Effects of Larval Density and Feeding Rates on the Bioconversion of Vegetable Waste Using Black Soldier Fly Larvae *Hermetia illucens* (L.), (Diptera: Stratiomyidae). *Waste Biomass Valorization* **2015**, *6*, 1059–1065. [CrossRef]
22. Barragan-Fonseca, K.B.; Dicke, M.; Van Loon, J.J.A. Influence of Larval Density and Dietary Nutrient Concentration on Performance, Body Protein, and Fat Contents of Black Soldier Fly Larvae (*Hermetia Illucens*). *Entomol. Exp. Appl.* **2018**, *166*, 761–770. [CrossRef] [PubMed]
23. Yakti, W.; Schulz, S.; Marten, V.; Mewis, I.; Padmanabha, M.; Hempel, A.-J.; Kobelski, A.; Streif, S.; Ulrichs, C. The Effect of Rearing Scale and Density on the Growth and Nutrient Composition of *Hermetia illucens* (L.) (Diptera: Stratiomyidae) Larvae. *Sustainability* **2022**, *14*, 1772. [CrossRef]
24. Amrul, N.F.; Kabir Ahmad, I.; Ahmad Basri, N.E.; Suja, F.; Abdul Jalil, N.A.; Azman, N.A. A Review of Organic Waste Treatment Using Black Soldier Fly (*Hermetia illucens*). *Sustainability* **2022**, *14*, 4565. [CrossRef]
25. Grau, M.G.P.; Dortmans, B.M.A.; Egger, J.; Virard, G.; Zurbrügg, C. Modelling the Financial Viability of Centralised and Decentralised Black Soldier Fly Larvae Waste Processing Units in Surabaya, Indonesia. *J. Insects Food Feed* **2023**, *9*, 303–316. [CrossRef]
26. Anyega, A.O.; Korir, N.K.; Beesigamukama, D.; Changeh, G.J.; Nkoba, K.; Subramanian, S.; Van Loon, J.J.A.; Dicke, M.; Tanga, C.M. Black Soldier Fly-Composted Organic Fertilizer Enhances Growth, Yield, and Nutrient Quality of Three Key Vegetable Crops in Sub-Saharan Africa. *Front. Plant Sci.* **2021**, *12*, 680312. [CrossRef]
27. Beesigamukama, D.; Tanga, C.M.; Sevgan, S.; Ekesi, S.; Kelemu, S. Waste to Value: Global Perspective on the Impact of Entomocomposting on Environmental Health, Greenhouse Gas Mitigation and Soil Bioremediation. *Sci. Total Environ.* **2023**, *902*, 166067. [CrossRef]
28. Menino, R.; Felizes, F.; Castelo-Branco, M.A.; Fareleira, P.; Moreira, O.; Nunes, R.; Murta, D. Agricultural Value of Black Soldier Fly Larvae Frass as Organic Fertilizer on Ryegrass. *Heliyon* **2021**, *7*, e05855. [CrossRef] [PubMed]
29. Setti, L.; Francia, E.; Pulvirenti, A.; Gigliano, S.; Zaccardelli, M.; Pane, C.; Caradonia, F.; Bortolini, S.; Maistrello, L.; Ronga, D. Use of Black Soldier Fly (*Hermetia illucens* (L.), Diptera: Stratiomyidae) Larvae Processing Residue in Peat-Based Growing Media. *Waste Manag.* **2019**, *95*, 278–288. [CrossRef]
30. Miranda, C.D.; Cammack, J.A.; Tomberlin, J.K. Mass Production of the Black Soldier Fly, *Hermetia illucens* (L.), (Diptera: Stratiomyidae) Reared on Three Manure Types. *Animals* **2020**, *10*, 1243. [CrossRef]
31. Liland, N.S.; Biancarosa, I.; Araujo, P.; Biemans, D.; Bruckner, C.G.; Waagbø, R.; Torstensen, B.E.; Lock, E.-J. Modulation of Nutrient Composition of Black Soldier Fly (*Hermetia Illucens*) Larvae by Feeding Seaweed-Enriched Media. *PLoS ONE* **2017**, *12*, e0183188. [CrossRef]
32. Gligorescu, A.; Fischer, C.H.; Larsen, P.F.; Nørgaard, J.V.; Heckman, L.-H.L. Production and Optimization of *Hermetia illucens* (L.) Larvae Reared on Food Waste and Utilized as Feed Ingredient. *Sustainability* **2020**, *12*, 9864. [CrossRef]
33. Gligorescu, A.; Macavei, L.I.; Larsen, B.F.; Markfoged, R.; Fischer, C.H.; Koch, J.D.; Jensen, K.; Lau Heckmann, L.-H.; Nørgaard, J.V.; Maistrello, L. Pilot Scale Production of *Hermetia illucens* (L.) Larvae and Frass Using Former Foodstuffs. *Clean. Eng. Technol.* **2022**, *10*, 100546. [CrossRef]
34. Lalander, C.; Diener, S.; Zurbrügg, C.; Vinnerås, B. Effects of Feedstock on Larval Development and Process Efficiency in Waste Treatment with Black Soldier Fly (*Hermetia illucens*). *J. Clean. Prod.* **2019**, *208*, 211–219. [CrossRef]
35. Deng, Y.-J.; Xiang, F.-M.; Tao, X.-H.; Jiang, C.-L.; Zhang, T.-Z.; Zhang, Z.-J. A Full-Scale Black Soldier Fly Larvae (*Hermetia Illucens*) Bioconversion System for Domestic Biodegradable Wastes to Resource. *Waste Manag. Res. J. Sustain. Circ. Econ.* **2023**, *41*, 143–154. [CrossRef]
36. Harnden, L.M.; Tomberlin, J.K. Effects of Temperature and Diet on Black Soldier Fly, *Hermetia illucens* (L.) (Diptera: Stratiomyidae), Development. *Forensic Sci. Int.* **2016**, *266*, 109–116. [CrossRef] [PubMed]
37. Sheppard, D.C.; Tomberlin, J.K.; Joyce, J.A.; Kiser, B.C.; Sumner, S.M. Rearing Methods for the Black Soldier Fly (Diptera: Stratiomyidae): Table 1. *J. Med. Entomol.* **2002**, *39*, 695–698. [CrossRef] [PubMed]
38. Hogsette, J.A. New Diets for Production of House Flies and Stable Flies (Diptera: Muscidae) in the Laboratory. *J. Econ. Entomol.* **1992**, *85*, 2291–2294. [CrossRef]
39. Georgescu, B.; Struti, D.; Papuc, T.; Ladosi, D.; Boaru, A. Body Weight Loss of Black Soldier Fly *Hermetia illucens* (Diptera: Stratiomyidae) during Development in Non-Feeding Stages: Implications for Egg Clutch Parameters. *Eur. J. Entomol.* **2020**, *117*, 216–225. [CrossRef]
40. Feldsine, P.; Abeyta, C.; Andrews, W.H. AOAC international Methods Committee Guidelines for Validation of Qualitative and Quantitative Food Microbiological Official Methods of Analysis. *J. AOAC Int.* **2002**, *85*, 1187–1200. [CrossRef]
41. Almeida, C.; Murta, D.; Nunes, R.; Baby, A.R.; Fernandes, Â.; Barros, L.; Rijo, P.; Rosado, C. Characterization of Lipid Extracts from the *Hermetia illucens* Larvae and Their Bioactivities for Potential Use as Pharmaceutical and Cosmetic Ingredients. *Heliyon* **2022**, *8*, e09455. [CrossRef]
42. *UNI EN ISO 6492*; Animal Feeding Stuffs—Determination of Fat Content. International Organization for Standardization: Geneva, Switzerland, 1999.

43. *UNI EN ISO 5983-1*; Animal Feeding Stuffs—Determination of Content and Calculation of Crude Protein Content—Part 1: Kjeldahl Method. International Organization for Standardization: Geneva, Switzerland, 2005.
44. Janssen, R.H.; Vincken, J.-P.; Van Den Broek, L.A.M.; Fogliano, V.; Lakemond, C.M.M. Nitrogen-to-Protein Conversion Factors for Three Edible Insects: *Tenebrio molitor*, *Alphitobius diaperinus*, and *Hermetia illucens*. *J. Agric. Food Chem.* **2017**, *65*, 2275–2278. [CrossRef]
45. *UNI EN ISO 18134-1*; Solid Biofuels—Determination of Moisture Content—Oven Dry Method—Part 1: Total Moisture—Reference Method. International Organization for Standardization: Geneva, Switzerland, 2015.
46. *UNI EN ISO 16948*; Solid Biofuels—Determination of Total Content of Carbon, Hydrogen and Nitrogen. International Organization for Standardization: Geneva, Switzerland, 2015.
47. Bascietto, M.; Santangelo, E.; Beni, C. Spatial Variations of Vegetation Index from Remote Sensing Linked to Soil Colloidal Status. *Land* **2021**, *10*, 80. [CrossRef]
48. Regione Emilia-Romagna. Pomodoro Da Industria. Parte Agronomica. Available online: http://agricoltura.regione.emilia-romagna.it/Produzioni-Agroalimentari/Temi/Bio-Agro-Climambiente/Agricoltura-Integrata/Disciplinari-Produzioneintegrata-Vegetale/Collezione-Dpi/2019/Orticole-2019 (accessed on 27 March 2024).
49. SIARL Lazio. Available online: https://www.siarl-lazio.it/E1_2.Asp (accessed on 27 March 2024).
50. Hammer, Ø.; Harper, D.A.T.; Ryan, P.D. PAST: Paleontological Statistics Software Package for Education and Data Analysis. *Palaeontol. Electron.* **2021**, *4*, 1.
51. Sripontan, Y.; Chiu, C.-I.; Tanansathaporn, S.; Leasen, K.; Manlong, K. Modeling the Growth of Black Soldier Fly *Hermetia illucens* (Diptera: Stratiomyidae): An Approach to Evaluate Diet Quality. *J. Econ. Entomol.* **2020**, *113*, 742–751. [CrossRef]
52. Oonincx, D.G.A.B.; Van Huis, A.; Van Loon, J.J.A. Nutrient Utilisation by Black Soldier Flies Fed with Chicken, Pig, or Cow Manure. *J. Insects Food Feed* **2015**, *1*, 131–140. [CrossRef]
53. Spranghers, T.; Ottoboni, M.; Klootwijk, C.; Ovyn, A.; Deboosere, S.; De Meulenaer, B.; Michiels, J.; Eeckhout, M.; De Clercq, P.; De Smet, S. Nutritional Composition of Black Soldier Fly (*Hermetia illucens*) Prepupae Reared on Different Organic Waste Substrates. *J. Sci. Food Agric.* **2017**, *97*, 2594–2600. [CrossRef] [PubMed]
54. Eggink, K.M.; Lund, I.; Pedersen, P.B.; Hansen, B.W.; Dalsgaard, J. Biowaste and By-Products as Rearing Substrates for Black Soldier Fly (*Hermetia illucens*) Larvae: Effects on Larval Body Composition and Performance. *PLoS ONE* **2022**, *17*, e0275213. [CrossRef] [PubMed]
55. Jucker, C.; Erba, D.; Leonardi, M.G.; Lupi, D.; Savoldelli, S. Assessment of Vegetable and Fruit Substrates as Potential Rearing Media for *Hermetia illucens* (Diptera: Stratiomyidae) Larvae. *Environ. Entomol.* **2017**, *46*, 1415–1423. [CrossRef]
56. Parry, N.J.; Weldon, C.W. Bioconversion of Poultry Manure and Vegetable Waste Mixes by *Hermetia illucens* (Diptera: Stratiomyidae). *J. Appl. Entomol.* **2023**, *147*, 279–288. [CrossRef]
57. Nguyen, T.T.X.; Tomberlin, J.K.; Vanlaerhoven, S. Ability of Black Soldier Fly (Diptera: Stratiomyidae) Larvae to Recycle Food Waste. *Environ. Entomol.* **2015**, *44*, 406–410. [CrossRef]
58. Parodi, A.; De Boer, I.J.M.; Gerrits, W.J.J.; Van Loon, J.J.A.; Heetkamp, M.J.W.; Van Schelt, J.; Bolhuis, J.E.; Van Zanten, H.H.E. Bioconversion Efficiencies, Greenhouse Gas and Ammonia Emissions during Black Soldier Fly Rearing—A Mass Balance Approach. *J. Clean. Prod.* **2020**, *271*, 122488. [CrossRef]
59. Sørensen, J.G.; Addison, M.F.; Terblanche, J.S. Mass-Rearing of Insects for Pest Management: Challenges, Synergies and Advances from Evolutionary Physiology. *Crop Prot.* **2012**, *38*, 87–94. [CrossRef]
60. Soma, D.D.; Maïga, H.; Mamai, W.; Bimbile-Somda, N.S.; Venter, N.; Ali, A.B.; Yamada, H.; Diabaté, A.; Fournet, F.; Ouédraogo, G.A.; et al. Does Mosquito Mass-Rearing Produce an Inferior Mosquito? *Malar. J.* **2017**, *16*, 357. [CrossRef]
61. Caruso, D.; Devic, E.; Subamia, I.W.; Talamond, P.; Baras, E. *Technical Handbook of Domestication and Production of Diptera Black Soldier Fly (BSF), Hermetia illucens, Stratiomyidae*; IPB Press: Bogor, Indonesia, 2014; pp. 65–76. ISBN 978-979-493-610-8.
62. Arabzadeh, G.; Delisle-Houde, M.; Tweddell, R.J.; Deschamps, M.-H.; Dorais, M.; Lebeuf, Y.; Derome, N.; Vandenberg, G. Diet Composition Influences Growth Performance, Bioconversion of Black Soldier Fly Larvae: Agronomic Value and In Vitro Biofungicidal Activity of Derived Frass. *Agronomy* **2022**, *12*, 1765. [CrossRef]
63. Meneguz, M.; Schiavone, A.; Gai, F.; Dama, A.; Lussiana, C.; Renna, M.; Gasco, L. Effect of Rearing Substrate on Growth Performance, Waste Reduction Efficiency and Chemical Composition of Black Soldier Fly (*Hermetia illucens*) Larvae. *J. Sci. Food Agric.* **2018**, *98*, 5776–5784. [CrossRef] [PubMed]
64. Venkatesh, K.; Morrison, P.E. Studies of Weight Changes and Amount of Foof Ingested by the Stable Fly, *Stomoxys calcitrans* (Diptera: Muscidae). *Can. Entomol.* **1980**, *112*, 141–149. [CrossRef]
65. Scala, A.; Cammack, J.A.; Salvia, R.; Scieuzo, C.; Franco, A.; Bufo, S.A.; Tomberlin, J.K.; Falabella, P. Rearing Substrate Impacts Growth and Macronutrient Composition of *Hermetia illucens* (L.) (Diptera: Stratiomyidae) Larvae Produced at an Industrial Scale. *Sci. Rep.* **2020**, *10*, 19448. [CrossRef]
66. Tschirner, M.; Simon, A. Influence of Different Growing Substrates and Processing on the Nutrient Composition of Black Soldier Fly Larvae Destined for Animal Feed. *J. Insects Food Feed* **2015**, *1*, 249–259. [CrossRef]
67. Yang, F.; Tomberlin, J.K. Comparing Selected Life-History Traits of Black Soldier Fly (Diptera: Stratiomyidae) Larvae Produced in Industrial and Bench-Top-Sized Containers. *J. Insect Sci.* **2020**, *20*, 25. [CrossRef]

68. Elissen, H.; van der Weide, R.; Gollenbeek, L. *Effects of Black Soldier Fly Frass on Plant and Soil Characteristics: A Literature Overview*; Report WPR-996; Wageningen Plant Research: Wageningen, The Netherlands, 2023; pp. 1–23.
69. Castellini, M.; Diacono, M.; Preite, A.; Montemurro, F. Short- and Medium-Term Effects of On-Farm Compost Addition on the Physical and Hydraulic Properties of a Clay Soil. *Agronomy* **2022**, *12*, 1446. [CrossRef]

MDPI AG
Grosspeteranlage 5
4052 Basel
Switzerland
Tel.: +41 61 683 77 34

Agriculture Editorial Office
E-mail: agriculture@mdpi.com
www.mdpi.com/journal/agriculture

www.ingramcontent.com/pod-product-compliance
Lightning Source LLC
Chambersburg PA
CBHW062041220926
51850CB00010B/2631
9783725856473